Engineering Economics and Costing

Dr. Prasun Bhattacharjee

DEDICATION

To the pioneers of progress and the architects of efficiency, this book is dedicated to those who engineer the future.

To the curious minds that navigate the intricate balance between cost and innovation, may this book be a compass guiding you through the landscapes of Engineering Economics and Costing.

In honor of the diligent learners and aspiring economists who decipher the language of numbers, may these pages empower you with insights that transcend equations.

To the relentless pursuit of optimal solutions, and the recognition that every decision has a cost, this work is dedicated.

May this book serve as a companion to those who seek to unravel the complexities of financial landscapes, and may the principles outlined herein foster a deeper understanding of the economic foundations that underpin progress.

With gratitude to the community of thinkers and practitioners who shape the discourse in Engineering Economics and Costing, this dedication acknowledges the collaborative spirit that propels our shared journey.

To the educators who inspire, the students who question, and the professionals who implement, may this book contribute to the ongoing dialogue that defines our ever-evolving field.

In acknowledgment of the challenges surmounted, the lessons learned, and the collective commitment to excellence, this dedication stands as a tribute to the spirit of innovation that thrives within the realm of Engineering Economics and Costing.

CONTENTS

Contents

ACKNOWLEDGMENTS

In the creation of this comprehensive exploration into the intricate world of "Engineering Economics and Costing," I am deeply grateful for the invaluable contributions and support from individuals whose wisdom and guidance have enriched this endeavor.

First and foremost, I extend my sincere appreciation to Prof. Amit Karmakar, the Head of the Department of Mechanical Engineering at Jadavpur University. His unwavering encouragement and scholarly insights have been instrumental in shaping the foundation of this book. His dedication to academic excellence and commitment to fostering a culture of learning have inspired me throughout this journey.

I extend my heartfelt thanks to Dr. Somenath Bhattacharya, Associate Professor at Jadavpur University, whose scholarly expertise and thoughtful critiques have significantly enhanced the quality of this work. His insightful suggestions and rigorous intellectual engagement have been invaluable in refining the content and ensuring its academic rigor.

My gratitude extends to Dr. Rabin Kr. Jana, Associate Professor at the Indian Institute of Management, Raipur, for his thoughtful contributions and perspectives that have added depth and breadth to the discussions within this book. His expertise in the field of management has brought a unique and complementary dimension to the exploration of economic principles.

I express my deepest gratitude to Swami Dheyananda ji Maharaj, the Officer-in-Charge of Ramakrishna Mission Shilpapitha, for his spiritual guidance and unwavering support throughout this journey. His wisdom and encouragement have been a source of inspiration, providing a holistic perspective that transcends the realms of academia.

To all those whose names may not appear in these pages but whose influence is deeply woven into the fabric of this work, I extend my sincere appreciation. Your collective efforts and contributions have left an indelible mark on the pages of "Engineering Economics and Costing."

— Dr. Prasun Bhattacharjee.

1 INTRODUCTION: ENGINEERING ECONOMICS AND ITS IMPORTANCE

The intersection of engineering and economics represents a dynamic and symbiotic relationship that plays a pivotal role in shaping the world we live in. While engineering is traditionally associated with the design and creation of innovative solutions to technical challenges, economics provides the framework for understanding the allocation of resources and the impact of these solutions on society. This chapter explores the intricate nexus between engineering and economics, delving into the ways in which these two disciplines inform and enrich each other. Engineering Economics, at its essence, is the application of economic principles to the decision-making process within the realm of engineering. It involves the systematic analysis of costs and benefits associated with engineering projects, helping practitioners make informed choices that optimize resources. The fundamental goal is to achieve efficiency and effectiveness in the utilization of resources while considering the economic viability of engineering endeavors. Engineering projects, whether they involve the construction of infrastructure, the development of new technologies, or the design of

1

manufacturing processes, are resource-intensive undertakings. Economic decision-making in engineering involves evaluating the costs and benefits associated with these projects to determine their feasibility and potential return on investment. This process requires a deep understanding of economic principles and methodologies.

1.1 Time Value of Money

One of the key concepts in engineering economics is the time value of money. Engineers must consider the fact that a sum of money today has a different value than the same sum in the future. This principle underlines the importance of discounting future cash flows and considering the present value of costs and benefits when making investment decisions. Realistic examples and case studies can illustrate how this concept impacts project evaluation and decision-making.

1.2 Resource Allocation and Opportunity Cost

The nexus between engineering and economics becomes particularly evident when examining resource allocation and opportunity cost. Engineers often face constraints in terms of budget, time, and available materials. Economics provides the analytical tools to evaluate alternative uses of resources, highlighting the concept of opportunity cost—the value of the next best alternative foregone when a decision is made. Balancing competing needs and making choices that maximize societal welfare become critical considerations in this context.

1.3 Cost-Benefit Analysis in Engineering

Cost-benefit analysis (CBA) is a fundamental tool in engineering economics that quantifies and compares the costs and benefits of a project. By assigning monetary values to both positive and negative outcomes, engineers can objectively assess whether a project is economically viable. This process involves not only financial considerations but also factors such as environmental impact, societal benefits, and risk analysis. Realistic examples can demonstrate how CBA is applied to make decisions that align with economic efficiency.

1.4 Economic Considerations in Project Management

Project management in engineering involves coordinating a multitude of activities, resources, and stakeholders. Economics provides a lens through which project managers can evaluate the economic feasibility of various project components, allocate resources efficiently, and ensure that project goals align with broader economic objectives. This includes considerations of economies of scale, risk management, and the overall financial health of a project.

1.5 Innovation and Economic Growth

The nexus between engineering and economics is a catalyst for innovation and economic growth. Technological advancements, driven by engineering breakthroughs, often lead to increased productivity and efficiency. This, in turn, contributes to economic growth by creating new industries, generating employment opportunities, and fostering a climate of innovation. Case studies can illuminate how iconic inventions, from the steam engine to the internet, have transformed economies and societies.

1.6 Regulatory and Policy Implications

The interplay between engineering and economics is evident in the regulatory and policy landscape. Governments and regulatory bodies formulate policies that impact engineering practices, addressing issues such as environmental sustainability, safety standards, and fair competition. The economic principles of market regulation, externalities, and public goods play a central role in shaping policies that aim to balance societal welfare with the imperatives of technological progress.

1.7 Globalization and Economic Integration

In an interconnected world, globalization further intertwines engineering and economics. International trade, the exchange of technological know-how, and collaborative research efforts transcend geographical boundaries. This global perspective requires engineers to consider economic implications on a broader scale, understanding how decisions made in one part of the world can have ripple effects across economies.

1.8 Sustainable Engineering Economics

Sustainability is an increasingly vital consideration in engineering economics. The nexus between engineering and environmental economics explores how sustainable practices can be integrated into engineering decision-making. Life cycle analysis, eco-design principles, and considerations of social and environmental impact contribute to a holistic approach that aligns engineering activities with long-term economic and ecological sustainability.

1.9 Challenges and Ethical Considerations

The nexus of engineering and economics is not without its challenges and ethical considerations. Engineers often grapple with dilemmas related to resource exploitation, social equity, and the unintended consequences of technological innovations. A nuanced understanding of economic ethics is essential for navigating these complexities, ensuring that engineering solutions contribute positively to societal well-being.

1.10 Conclusion

In conclusion, the nexus of engineering and economics represents a dynamic and multifaceted relationship that shapes the trajectory of human progress. From the conceptualization of projects to their execution and the societal impacts they generate, the interplay between these two disciplines is profound. As we navigate through this nexus, it becomes clear that a holistic understanding of engineering economics is essential for practitioners and decision-makers in the field. This chapter serves as a foundational exploration, setting the stage for a deeper dive into specific aspects of engineering economics in the chapters that follow. The journey continues, unveiling the intricacies of a discipline that bridges the realms of innovation and economic reasoning.

Suggested Questions with Solutions

1. Question: Why is the integration of economic principles crucial in the field of engineering? Provide a comprehensive discussion on how Engineering Economics enhances decision-making in engineering projects.

Solution: The integration of economic principles in engineering is essential as it enables engineers to make decisions that are not only technically sound but also economically viable. Engineering Economics provides a systematic approach to analyzing costs, benefits, and risks associated with projects, ensuring optimal resource allocation and overall project success. It facilitates decision-making by considering factors such as the time value of money, opportunity costs, and ethical considerations, leading to well-informed and sustainable engineering solutions.

2. Question: Explore the impact of the time value of money on project evaluation in Engineering Economics. How does considering the time value of money influence long-term planning and decision-making in engineering projects?

Solution: The time value of money significantly influences project evaluation by recognizing that the value of money changes over time. Considering the time value of money is crucial in long-term planning as it affects the present and future worth of cash flows, influencing investment decisions and project feasibility. Engineers must account for interest rates, inflation, and discounting when evaluating the economic viability of projects to ensure realistic financial planning and sustainable outcomes.

3. Question: Discuss the role of opportunity cost in engineering decision-making. How does the concept of opportunity cost guide engineers in resource allocation and project prioritization? Provide examples to illustrate your points.

Solution: Opportunity cost plays a vital role in engineering decision-making by highlighting the value of the next best alternative forgone. Engineers often face choices in allocating resources, and considering opportunity costs helps in making informed decisions. For example, in a construction project, opting for a more expensive but durable material might involve the opportunity cost of not choosing a less expensive alternative with a shorter lifespan. By understanding opportunity costs, engineers can prioritize projects and resources effectively.

4. Question: Explore the ethical dimensions of Engineering Economics. How do ethical considerations influence decision-making in engineering projects, and what role do engineers play in ensuring ethical practices in their work?

Solution: Ethical considerations in Engineering Economics encompass a range of factors, including fair resource distribution, environmental impact, and social responsibility. Engineers play a crucial role in ensuring ethical practices by making decisions that prioritize societal well-being, environmental sustainability, and transparency. For instance, in the design of infrastructure projects, ethical considerations may involve minimizing negative impacts on local communities and ecosystems, demonstrating a commitment to responsible engineering.

5. Question: Evaluate the global perspectives in Engineering Economics. How does globalization impact engineering decision-making, and what challenges and opportunities arise for engineers in a globalized context? Provide examples to illustrate your points.

Solution: Global perspectives in Engineering Economics arise from the interconnected nature of economies and technological advancements. Globalization impacts engineering decision-making by introducing diverse perspectives, cultural considerations, and market dynamics. Engineers working on international projects must navigate challenges such as varying regulatory frameworks, cultural differences, and environmental standards. For example, a multinational corporation implementing a manufacturing process in different countries must consider local economic conditions, regulations, and community expectations. Globalization presents both challenges and opportunities for engineers to contribute to sustainable and inclusive development on a global scale.

2 WANT ACTIVITY – SATISFACTION OF WANTS

𝕳uman existence is a perpetual journey marked by desires, needs, and the pursuit of satisfaction. This intricate dance of wants and their fulfillment, often referred to as "want activity," is a fundamental aspect of the human experience. It encompasses a vast spectrum of desires, ranging from basic necessities to complex aspirations, and is shaped by economic, social, and psychological factors. This exploration delves into the dynamic interplay of want activity and the nuanced ways in which individuals and societies seek to satisfy their myriad wants.

2.1 Defining Wants

Wants represent a diverse array of desires that individuals seek to fulfill. From the fundamental needs for food, shelter, and clothing to more intricate aspirations like education, cultural experiences, and self-actualization, wants form the mosaic of human yearning. Understanding the nature of wants is essential for unraveling the complexities of want activity.

2.2 Dynamic and Subjective Nature

Want activity is dynamic, fluid, and highly subjective. The hierarchy of needs, as proposed by psychologist Abraham Maslow, illustrates that wants evolve as individuals progress through various stages of life. What satisfies a want for one person may differ significantly for another, adding layers of complexity to the pursuit of satisfaction.

2.3 Satisfaction of Wants: Economic Dimensions

2.3.1 Resource Allocation and Scarcity

The satisfaction of wants is intricately tied to economic principles, particularly the fundamental concept of scarcity. With limited resources and unlimited wants, individuals and societies face the challenge of allocating resources efficiently. The choices made in this process determine the satisfaction of wants and give rise to economic activities.

2.3.2 Opportunity Cost

Want activity involves decision-making, and every decision comes with an opportunity cost—the value of the next best alternative foregone. For example, choosing to invest time in one activity may mean sacrificing the opportunity to engage in another. This concept is integral to understanding the trade-offs inherent in the pursuit of satisfaction.

2.3.3 Consumer Behavior and Market Dynamics

The study of consumer behavior provides insights into the intricacies of want satisfaction. Psychological, social, and economic factors influence how individuals make choices regarding the purchase and use of goods and services. In the marketplace, the forces of supply and demand, along with advertising and branding, play a crucial role in shaping want activity and influencing satisfaction.

2.4 Innovation and Want Fulfillment

2.4.1 Driven by Unmet Needs

Innovation often arises from the desire to fulfill unmet needs. Engineers, scientists, and entrepreneurs respond to challenges and aspirations by creating new products and services. The cycle of want activity stimulates innovation, leading to advancements that cater to evolving desires and preferences.

2.4.2 Technological Advancements

The satisfaction of wants is closely linked to technological progress. Advancements in technology not only create new wants but also provide innovative solutions to existing desires. For example, the advent of smartphones addressed the want for instant communication, while advancements in medical technology cater to the desire for improved health and well-being.

2.5 Challenges in Want Satisfaction: Balancing Act and Ethical Considerations

2.5.1 Scarcity and Opportunity Cost

The fundamental challenge in want satisfaction lies in the scarcity of resources. Balancing competing wants and allocating resources judiciously require individuals and societies to confront the opportunity cost of their choices. Understanding and navigating scarcity form a central aspect of the want satisfaction journey.

2.3.2 Societal and Environmental Impact

Want activity, when unchecked, can have profound societal and environmental consequences. Unsustainable consumption patterns contribute to resource depletion, environmental degradation, and social inequality. Ethical considerations come into play as individuals and societies grapple with the broader impact of their wants on the well-being of the planet and its inhabitants.

2.6 Ethical Dimensions of Want Activity

2.6.1 Social Equity

Ethical considerations in want activity extend to social equity. Ensuring fair access to resources and opportunities becomes imperative to address disparities and promote inclusivity. Ethical choices in want satisfaction involve recognizing and addressing societal inequalities, working towards a more just distribution of resources.

2.6.2 Responsible Consumption

Ethical want activity embraces responsible consumption. This involves conscious choices that consider the environmental and social impact of consumption patterns. From sustainable sourcing to eco-friendly practices, responsible consumption aligns individual wants with broader ethical and environmental goals.

2.7 Conclusion: Navigating the Intricacies of Want Activity

Want activity, with its rich tapestry of desires and the pursuit of satisfaction, weaves through the fabric of human existence. From the economic considerations of scarcity and resource allocation to the ethical dimensions of social equity and responsible consumption, the dynamics of want activity are complex and multifaceted. Understanding and navigating want activity require a holistic approach—one that recognizes the subjective nature of desires, appreciates the economic principles at play, and embraces ethical considerations. As individuals and societies evolve, so too does the dance of want activity, influencing the choices we make, the innovations we create, and the impact we have on the world. In this intricate dance of wants and their satisfaction, individuals are invited to reflect on their own desires, choices, and the ethical dimensions of their want activity. By doing so, the journey of want satisfaction becomes not only a personal exploration but also a collective endeavor towards a more balanced, equitable, and sustainable world.

Suggested Questions with Solutions

1. Question: Explore the psychological and economic factors influencing want activity. How do these factors shape individual preferences and contribute to the broader dynamics of economic decision-making?

Solution: Want activity is intricately linked to both psychological and economic factors. Psychologically, individual preferences are shaped by cultural influences, personal experiences, and social norms. Economically, factors such as income, price levels, and market dynamics play a crucial role. The interaction of these psychological and economic factors contributes to the complexity of want activity, influencing the choices individuals make and the impact on broader economic decision-making.

2. Question: Discuss the role of innovation in satisfying evolving wants. How do technological advancements contribute to the creation of new desires, and in what ways do they address existing ones?

Solution: Innovation plays a pivotal role in the satisfaction of wants. Technological advancements often lead to the creation of new desires by introducing novel products and services. For instance, the emergence of smartphones addressed the desire for instant communication. Simultaneously, innovation addresses existing wants by providing more efficient and advanced solutions. Understanding this dynamic relationship between innovation and want satisfaction is crucial for anticipating and adapting to evolving consumer needs.

3. Question: Analyze the ethical considerations involved in want activity. How can responsible consumption practices align individual desires with broader environmental and social goals?

Solution: Ethical considerations in want activity revolve around responsible consumption. Individuals can align their desires with broader environmental and social goals by making conscious and ethical choices. This involves considering the environmental impact of products, supporting sustainable practices, and ensuring fair and ethical sourcing. Responsible consumption empowers individuals to contribute positively to societal well-being and environmental sustainability.

4. Question: Explore the global dimensions of want activity. How does globalization influence the nature of desires and the strategies employed for their satisfaction on a global scale?

Solution: Globalization has a profound impact on want activity, influencing the nature of desires and the strategies employed for their satisfaction. Cultural exchange, international trade, and technological interconnectedness contribute to the globalization of desires. Understanding how global dynamics shape want activity is essential for businesses, policymakers, and individuals navigating a world where desires are increasingly interconnected and transcendent of geographical boundaries.

5. Question: Evaluate the role of scarcity and opportunity cost in want satisfaction. How do these economic principles shape decision-making at both individual and societal levels?

Solution: Scarcity and opportunity cost are fundamental economic principles that shape want satisfaction. Scarcity highlights the limited availability of resources, prompting individuals and societies to make choices. Opportunity cost represents the value of the next best alternative foregone in these choices. Navigating scarcity and opportunity cost involves weighing trade-offs and making decisions that optimize resource allocation. Understanding these economic principles is crucial for making informed and efficient choices in the pursuit of satisfying wants.

3 RESOURCE PLANNING AND DISTRIBUTION IN ECONOMIC SYSTEMS: A COMPREHENSIVE ANALYSIS OF LAISSEZ-FAIRE

$\mathfrak{R}$esource planning and distribution are pivotal elements in the functioning of economic systems, influencing the allocation of resources to meet the needs and wants of society. Among the various approaches to economic systems, laissez-faire stands out as a philosophy that emphasizes minimal government intervention in economic affairs. In this exploration, we delve into the intricate dynamics of resource planning and distribution within laissez-faire economic systems, examining the principles, advantages, challenges, and societal implications.

3.1 Understanding Laissez-Faire Economics

Laissez-faire economics, rooted in classical liberal economic thought, embodies the principle of minimal government intervention in economic affairs. The term, derived from French and roughly translated as "let do" or "leave alone," signifies an economic philosophy that champions the belief in free markets as the most efficient allocators of resources. According to laissez-faire economics, individuals pursuing their self-interest in a competitive marketplace will naturally contribute to overall economic prosperity. This approach emphasizes limited government interference, advocating for the removal of barriers to entry, free competition, and the protection of private property rights. Proponents argue that the invisible hand of the market, as coined by economist Adam Smith, guides economic activities for the collective good without the need for centralized control. While laissez-faire economics has its merits in fostering innovation and efficiency, critics raise concerns about potential market failures, income inequality, and the need for some level of regulation to address externalities and ensure social welfare. The philosophy has played a significant role in shaping economic policies and debates throughout history, influencing the trajectory of capitalism and government involvement in the market.

3.1.1 Definition and Origins
Laissez-faire, a French term meaning "let it be" or "leave it alone," represents an economic philosophy advocating for minimal government interference in market activities. Originating in the 18th-century writings of economists like Adam Smith, laissez-faire economics asserts that markets, if left to operate freely, will efficiently allocate resources and generate optimal outcomes.

3.1.2. Key Principles
The core principles of laissez-faire economics include the belief in the self-regulating nature of markets, the invisible hand mechanism, and the idea that individual pursuit of self-interest leads to collective societal benefits. In a laissez-faire system, the government's role is limited to safeguarding property rights and maintaining a legal framework that ensures fair competition.

3.2 Resource Planning in Laissez-Faire Systems
Resource planning in laissez-faire systems operates under the premise that, in a free-market environment, individual actors pursuing their self-interest will collectively optimize the allocation of resources. In laissez-faire economics, the emphasis is on minimal government intervention, allowing market forces to dictate resource distribution

based on supply and demand dynamics. This approach contends that a decentralized system, where individuals and businesses make independent decisions in response to market signals, leads to the most efficient utilization of resources. Resource planning, therefore, becomes a spontaneous and decentralized process guided by the invisible hand of the market, as articulated by classical economists like Adam Smith. In laissez-faire systems, entrepreneurs and businesses dynamically respond to changing market conditions, adjusting their production and resource utilization strategies based on consumer preferences and competitive pressures. While proponents argue that this decentralized resource planning fosters innovation, flexibility, and responsiveness, critics point to potential drawbacks, such as market failures, externalities, and the risk of unequal resource distribution that may necessitate some degree of regulatory intervention.

3.2.1 Market Mechanism

Laissez-faire relies on the market mechanism to plan and allocate resources. In a free-market system, prices act as signals, conveying information about scarcity and demand. The interaction of buyers and sellers in open markets determines the allocation of resources based on consumer preferences and the profitability of goods and services.

3.2.2 Supply and Demand Dynamics

The forces of supply and demand play a central role in resource planning within laissez-faire systems. Prices adjust based on the equilibrium of supply and demand, guiding producers and consumers to make decisions that balance the availability of resources with societal needs and wants.

3.2.3 Entrepreneurship and Innovation

Laissez-faire systems encourage entrepreneurship and innovation as drivers of resource planning. Entrepreneurs, motivated by the prospect of profit, identify opportunities to allocate resources more efficiently or create new products and services that cater to changing consumer demands.

3.3 Resource Distribution in Laissez-Faire Systems

Resource distribution in laissez-faire systems is fundamentally driven by the principles of free-market economics, where the forces of supply and demand autonomously determine how resources are allocated. In this system, individuals and businesses, acting in their self-interest, engage in voluntary exchanges based on market prices. The allocation of resources is guided by consumer preferences, competition, and the pursuit of profit. Those resources that are in high demand typically command higher prices, signaling producers to allocate more resources to meet that demand. Conversely, resources that are less sought after may see a reduction in allocation. The decentralized nature of laissez-faire systems means that resource distribution is dynamic and responsive to changing market conditions. Advocates argue that this approach fosters efficiency, innovation, and economic growth. However, critics point to potential drawbacks, such as income inequality, market failures, and the need for regulatory measures to address externalities and ensure fair competition. In laissez-faire systems, the market mechanism plays a central role in shaping the pattern of resource distribution.

3.3.1 Income Distribution

Laissez-faire systems are often associated with a certain pattern of income distribution. The philosophy contends that individuals, through their skills, efforts, and entrepreneurship, contribute to the creation of wealth. As a result, income distribution is expected to align with individual contributions to the economy.

3.3.2 Wealth Accumulation and Inequality

The pursuit of self-interest and the accumulation of wealth are integral aspects of laissez-faire economics. However, this can lead to concerns about income inequality. The concentration of wealth in the hands of a few individuals or entities may result in disparities in access to resources, opportunities, and a potential impact on social cohesion.

3.4 Advantages of Laissez-Faire Resource Planning

Laissez-faire resource planning comes with several advantages rooted in the principles of free-market economics. One of the key benefits is the inherent efficiency that arises from decentralized decision-making. In laissez-faire systems, individual actors, driven by self-interest, respond dynamically to market signals, optimizing the allocation of resources based on consumer demand and competition.

This flexibility allows for quick adjustments to changing conditions, fostering innovation and adaptability. Additionally, the competition inherent in free markets encourages cost efficiency and drives businesses to deliver goods and services of higher quality. Laissez-faire resource planning is often associated with economic growth and wealth creation, as it allows entrepreneurs and businesses to pursue opportunities without undue government interference. Proponents argue that this approach promotes individual freedoms, encourages entrepreneurship, and enhances overall economic prosperity. However, it is essential to recognize that while laissez-faire systems have notable advantages, they also require careful consideration of potential drawbacks, such as income inequality and the need for regulatory measures to address market failures and externalities.

3.4.1 Efficiency and Innovation

Laissez-faire systems are praised for their efficiency in resource allocation. The decentralized decision-making process allows for quick responses to changing conditions, fostering innovation and adaptability. Entrepreneurs are motivated to seek efficient solutions to maximize profits, driving technological advancements and productivity gains.

3.4.2 Individual Freedom and Choice

Laissez-faire emphasizes individual freedom and choice. In such systems, consumers have the autonomy to make purchasing decisions based on their preferences, and producers can respond to these preferences without undue interference. This freedom is considered a fundamental aspect of laissez-faire's appeal.

3.4.3. Market Signals and Information

Prices in a laissez-faire system act as signals, conveying information about scarcity and demand. This transparent pricing mechanism allows participants in the market to make informed decisions, guiding resource allocation in a way that reflects the preferences and needs of consumers.

3.5 Challenges and Criticisms

Laissez-faire economics, while praised for its emphasis on

individual freedoms and market efficiency, has faced significant challenges and criticisms. One of the primary concerns is the potential for market failures, where unfettered competition may lead to suboptimal outcomes. Critics argue that laissez-faire systems might not adequately address issues such as externalities, monopolies, and information asymmetry, necessitating government intervention to rectify these imbalances. Another criticism revolves around income inequality, as free-market dynamics can result in the concentration of wealth and opportunities in the hands of a few. Laissez-faire economics also faces challenges related to social and environmental considerations, where profit-maximizing behaviors may disregard broader societal or ecological concerns. The 2008 financial crisis and other economic downturns have fueled skepticism about the self-regulating nature of markets, prompting calls for a more balanced approach that incorporates appropriate regulations. While laissez-faire economics has undeniable strengths, these challenges underscore the importance of finding a nuanced and pragmatic balance between free-market principles and the need for effective governance to ensure fairness, stability, and social well-being.

3.5.1 Market Failures

One of the main criticisms of laissez-faire economics is its vulnerability to market failures. In situations where markets do not efficiently allocate resources, such as in the case of externalities, public goods, or monopolies, laissez-faire may result in suboptimal outcomes and social inefficiencies.

3.5.2 Income Inequality and Social Welfare

The accumulation of wealth and income inequality is a persistent challenge within laissez-faire systems. Critics argue that without intervention, these systems may fail to address issues of poverty and social welfare adequately. The unequal distribution of resources can lead to disparities in education, healthcare, and overall quality of life.

3.5.3 Short-Term Focus and Externalities

Laissez-faire systems, driven by profit motives, may prioritize short-term gains over long-term sustainability. Concerns arise regarding the potential neglect of environmental externalities and the depletion of natural resources in the pursuit of immediate economic benefits.

3.6 Societal Implications and Considerations

Laissez-faire economics carries significant societal implications and considerations that extend beyond the realm of economic theory. The philosophy's emphasis on limited government intervention and the pursuit of individual self-interest can shape the fabric of societies in various ways. Proponents argue that laissez-faire systems, by fostering free-market competition, stimulate economic growth and innovation, leading to overall prosperity. However, societal implications also include challenges such as income inequality, as the benefits of economic growth may not be evenly distributed. The laissez-faire approach may contribute to the commodification of various aspects of life, potentially neglecting social and environmental concerns in the relentless pursuit of profit. Moreover, the philosophy assumes a level playing field, but in reality, disparities in access to resources and opportunities can persist, affecting social mobility. Balancing the virtues of individual freedom and market efficiency with the need for social justice, environmental sustainability, and equitable resource distribution remains a complex challenge in considering the societal impact of laissez-faire economics. It underscores the importance of thoughtful policies and regulations to mitigate potential negative consequences and ensure a more inclusive and socially responsible economic framework.

3.6.1 Role of Government in Social Welfare

Laissez-faire systems prompt discussions about the role of government in ensuring social welfare. While the philosophy advocates for minimal intervention, there is a recognition that some level of government involvement may be necessary to address market failures and mitigate the negative social consequences of income inequality.

3.6.2. Balancing Freedom and Regulation

Achieving a balance between individual freedom and necessary regulation is a complex challenge in laissez-faire systems. Striking the right balance requires careful consideration of how regulatory measures can enhance market efficiency while addressing societal needs, preventing exploitation, and safeguarding public goods.

3.7 Conclusion: Navigating the Path of Laissez-Faire Economics

Resource planning and distribution within laissez-faire economic systems represent a dynamic and complex interplay of market forces, individual actions, and societal outcomes. The principles of minimal government intervention, decentralized decision-making, and reliance on market mechanisms define the laissez-faire approach. While it offers advantages in terms of efficiency, innovation, and individual freedom, challenges such as market failures and income inequality necessitate careful consideration. The path of laissez-faire economics involves navigating the delicate balance between individual aspirations and societal well-being. As economies evolve and global challenges emerge, the application of laissez-faire principles requires adaptation, ethical considerations, and a commitment to addressing the broader implications of resource planning and distribution on the fabric of society. In the ongoing discourse on economic systems, understanding the nuances of laissez-faire economics is essential for informed decision-making and shaping the future of resource allocation in a rapidly changing world.

Suggested Questions with Solutions

1. Question: Assess the core principles of laissez-faire economics in the context of resource planning. How does the philosophy rely on decentralized decision-making and market mechanisms to allocate resources efficiently?

Solution: Laissez-faire economics relies on the principles of minimal government intervention and decentralized decision-making. In terms of resource planning, the philosophy emphasizes the efficiency of market mechanisms. Decentralized decision-making allows individual actors in the market to respond dynamically to changes in supply and demand, ensuring that resources are allocated based on consumer preferences and profitability. This approach, rooted in the belief of the self-regulating nature of markets, forms the foundation of laissez-faire resource planning.

2. Question: Explore the role of supply and demand dynamics in resource distribution within a laissez-faire economic system. How do these forces influence pricing and guide the allocation of resources in a free-market environment?

Solution: In laissez-faire systems, supply and demand dynamics are fundamental to resource distribution. The forces of supply and demand interact to determine prices in a free-market environment. When demand exceeds supply, prices rise, signaling an opportunity for producers to allocate more resources to meet consumer needs. Conversely, when supply exceeds demand, prices fall, indicating a need for resource reallocation. The price mechanism guides resource distribution, ensuring that resources flow to where they are most valued in the market.

3. Question: Examine the advantages of laissez-faire resource planning, particularly in terms of efficiency and individual freedom. How does the philosophy encourage entrepreneurship and innovation as drivers of optimal resource allocation?

Solution: Laissez-faire resource planning offers several advantages. The philosophy is known for its efficiency in resource allocation, driven by the decentralized decision-making process. Entrepreneurs, motivated by the pursuit of profit, identify opportunities for more efficient resource allocation and drive innovation to meet changing consumer demands. The individual freedom inherent in laissez-faire systems empowers consumers to make choices based on their preferences, contributing to a dynamic and responsive market environment.

4. Question: Critically evaluate the challenges and criticisms associated with laissez-faire resource planning. How does the philosophy address concerns related to market failures, income inequality, and potential neglect of long-term sustainability?

Solution: Laissez-faire economics faces criticisms and challenges, including concerns about market failures and income inequality. Market failures, such as externalities and monopolies, may result in suboptimal outcomes. Income inequality can arise from the concentration of wealth. Additionally, laissez-faire systems may prioritize short-term gains over long-term sustainability. While advocates argue that the self-regulating nature of markets can address some of these challenges, critics call for careful consideration and potential government intervention to mitigate negative consequences.

5. Question: Discuss the societal implications and considerations associated with laissez-faire resource planning. How does the philosophy impact income distribution, social welfare, and the role of government in addressing societal needs?

Solution: Laissez-faire resource planning has societal implications, particularly in terms of income distribution and social welfare. The philosophy's emphasis on individual contributions to the economy may result in certain patterns of income distribution. However, concerns about social welfare, poverty, and inequality prompt discussions on the role of government. Striking a balance between individual freedom and necessary regulation becomes crucial, as governments may need to intervene to address market failures, ensure social welfare, and prevent exploitation.

4 FACTORS OF PRODUCTION: THE CONCEPT OF OPTIMUM AND LAWS OF RETURN

The concept of factors of production lies at the heart of economic theory, representing the fundamental building blocks necessary for the creation of goods and services within an economy. The four primary factors—land, labor, capital, and entrepreneurship—interact in complex ways, and their optimal utilization is critical for achieving maximum efficiency and output. In this comprehensive exploration, we delve into the concept of optimum in factors of production, examining how each factor contributes to the overall production process. Additionally, we explore the laws of return, fundamental principles that govern the relationship between inputs and outputs, shedding light on the dynamics of production efficiency.

4.1 Factors of Production

Factors of production refer to the various inputs required for the process of creating goods and services in an economic system. Traditionally, economists classify these factors into four categories: land, labor, capital, and entrepreneurship. Land encompasses all natural resources used in production, such as agricultural land, minerals, and water. Labor represents the human effort and skill involved in production, ranging from manual labor to intellectual contributions. Capital includes both physical capital (machinery, tools, buildings) and financial capital (money) used to facilitate production. Lastly, entrepreneurship embodies the innovative and managerial skills required to organize the other factors efficiently and take on the risks associated with business ventures. These factors operate in tandem, and their combination determines a society's ability to generate goods and services. The study and understanding of factors of production are central to economic theory, guiding discussions on resource allocation, productivity, and overall economic development.

4.1.1 Land

Land, in economic terms, encompasses the natural resources used in production. This includes the physical space, minerals, water bodies, forests, and other resources derived from the Earth. The concept of optimum land utilization involves considering various factors, such as soil quality, climate, and location, to determine the most efficient and productive use of this finite resource. For example, agricultural activities require a thorough understanding of soil fertility and climate patterns to optimize crop yields.

4.1.2 Labor

Labor constitutes the human input in the production process, encompassing physical and intellectual efforts. Achieving the optimum level of labor involves considerations such as the number of workers, their skills, motivation, and working conditions. Workforce management, training programs, and the overall well-being of employees contribute to the efficiency and effectiveness of labor in the production process. Striking the right balance in the utilization of human resources is crucial for achieving maximum output.

4.1.3 Capital

Capital represents the man-made resources employed in production, including machinery, equipment, and infrastructure. The concept of optimum capital utilization revolves around finding the right mix and level of investment to maximize output without unnecessary waste. Factors such as technological advancements, maintenance, and depreciation play a crucial role in determining the optimal use of capital in the production process. Technological innovation, in particular, can significantly impact the efficiency and productivity of capital-intensive industries.

4.1.4 Entrepreneurship

Entrepreneurship involves the ability to innovate, take risks, and organize the other factors of production. The entrepreneur is the driving force behind identifying opportunities, making strategic decisions, and navigating the challenges of the business environment. Optimum entrepreneurship requires effective leadership, risk management, and a keen understanding of market dynamics. Entrepreneurs play a pivotal role in coordinating the factors of production to create value and drive economic growth.

4.2 The Concept of Optimum in Factors of Production

The concept of optimum in factors of production refers to the ideal combination and utilization of the various inputs—land, labor, capital, and entrepreneurship—to achieve maximum efficiency and output in the production process. The goal is to find the most effective balance that maximizes output while minimizing costs. Achieving an optimum level involves determining the right mix of inputs, taking into consideration factors such as technology, market conditions, and resource availability. The concept recognizes that an imbalance or underutilization of any factor can lead to suboptimal results. Economies strive to reach this optimum point to ensure the most efficient use of resources and to enhance overall productivity. The pursuit of the optimum in factors of production is a fundamental aspect of economic decision-making, guiding businesses, policymakers, and entrepreneurs in their efforts to achieve sustainable growth and prosperity.

4.2.1 Optimum Land Utilization

Achieving optimum land utilization involves understanding the characteristics of the land and making decisions that maximize productivity. Different types of land may be suitable for specific purposes, such as agriculture, residential development, or industrial activities. For example, fertile land with access to water sources is ideal for agricultural cultivation, while land in urban areas may be better suited for industrial or residential development. The concept of optimum land utilization also extends to sustainable practices. As concerns about environmental conservation and climate change rise, optimizing land use involves balancing economic activities with the preservation of ecosystems. This may entail adopting practices that reduce soil erosion, promote biodiversity, and mitigate the impact of human activities on the land.

4.2.2 Optimum Labor Utilization

Optimizing labor utilization involves ensuring the right number of workers with the appropriate skills are employed to meet production demands. Workforce planning and management play a crucial role in achieving this balance. Underutilizing labor may lead to inefficiencies, while overutilizing it can result in burnout and decreased productivity. Moreover, the concept of optimum labor utilization extends beyond numerical considerations. It includes creating a conducive work environment, providing training opportunities, and fostering employee engagement. When workers are motivated, well-trained, and satisfied with their work conditions, they are likely to contribute more effectively to the production process.

4.2.3 Optimum Capital Utilization

Optimizing capital utilization requires careful consideration of the level and composition of capital investment. The goal is to strike a balance between investing enough to enhance productivity without overspending on unnecessary resources. Technological advancements play a significant role in determining the optimum level of capital investment, as new technologies can increase efficiency and reduce the need for certain types of capital. Additionally, the concept of optimum capital utilization involves maintaining and upgrading capital assets to ensure their longevity and effectiveness. Regular maintenance and periodic upgrades can prevent inefficiencies caused by equipment breakdowns and technological obsolescence.

4.2.4 Optimum Entrepreneurship

Optimizing entrepreneurship involves harnessing the innovative and organizational capabilities of entrepreneurs to drive business success. Successful entrepreneurs possess a vision for the future, the ability to identify and exploit market opportunities, and the resilience to navigate challenges. The concept of optimum entrepreneurship also includes effective decision-making, risk management, and the ability to adapt to changing market conditions. Encouraging entrepreneurship at both the individual and institutional levels contribute to economic dynamism. Policies that support the development of entrepreneurial skills, provide access to funding, and create a supportive business environment can enhance the overall entrepreneurial ecosystem.

4.3 Laws of Return

The Laws of Return, also known as the Laws of Production, are economic principles that describe the relationship between inputs and outputs in the production process. These laws are crucial in understanding how changes in factors of production impact overall output. The Law of Diminishing Returns is a fundamental concept stating that as one input is increased while other inputs are held constant, the marginal output (additional output gained) will eventually decrease. In simpler terms, there is a point at which adding more of a particular input yields progressively smaller increases in output. This law underscores the importance of balancing factors of production to optimize efficiency. On the other hand, the Law of Increasing Returns suggests that increasing certain inputs may result in a disproportionate increase in output, leading to higher levels of productivity. Both laws play a central role in economic decision-making, guiding producers and policymakers in resource allocation and production strategies. Understanding these laws helps to ensure that resources are utilized efficiently and that the production process aligns with economic goals and sustainability.

4.3.1 Law of Increasing Returns

The Law of Increasing Returns states that as additional units of one input are added to the production process while keeping other inputs constant, the total output increases at an increasing rate. This implies that the marginal product of the variable input is rising. The phenomenon of increasing returns often occurs in situations where specialization and division of labor play a significant role. For instance, in manufacturing processes, as more workers specialize in specific tasks, the overall productivity of the entire production system may increase. This law highlights the potential benefits of scaling up production and achieving economies of scale.

4.3.2 Law of Constant Returns

The Law of Constant Returns posits that as additional units of one input are added, the total output increases at a constant rate. This implies that the marginal product of the variable input remains constant. While constant returns are less common in the real world compared to increasing or diminishing returns, they are essential for understanding production dynamics in certain scenarios. Agricultural production, for example, may exhibit constant returns if the cultivation process remains unchanged and additional units of labor or capital are applied in a uniform manner. Understanding constant returns is crucial for making informed decisions about resource allocation and production strategies.

4.3.3 Law of Diminishing Returns

The Law of Diminishing Returns asserts that as more units of a variable input are added to the production process while keeping other inputs constant, the total output increases at a decreasing rate. In other words, the marginal product of the variable input declines. This law reflects the reality that, at a certain point, the efficiency gains from adding more of a particular input diminish. For instance, in agriculture, as more fertilizer is applied to a fixed area of land, there is initially an increase in crop yield. However, beyond a certain point, adding more fertilizer may lead to diminishing returns, as the soil becomes saturated, and additional fertilizer has a reduced impact on crop growth.

4.4 Practical Implications and Case Studies

Practical implications of economic theories, such as the Laws of Return, become evident through real-world applications and case studies. For instance, in agriculture, the Law of Diminishing Returns is observed when farmers apply more fertilizer to a fixed area of land.

Initially, yields may increase, but beyond a certain point, additional fertilizer may lead to diminishing returns as the soil becomes saturated. This understanding guides farmers to optimize fertilizer use for maximum crop yield. Conversely, in the technology sector, the Law of Increasing Returns is evident in network effects, where the value of a product or service increases as more users adopt it. For example, social media platforms and operating systems often experience increasing returns due to the positive feedback loop created by a growing user base. These practical implications emphasize the importance of strategic resource allocation, efficiency, and adapting production processes to maximize output in various industries. Case studies provide valuable insights into how economic principles are applied, helping businesses and policymakers make informed decisions for sustainable growth and development.

4.4.1 Agriculture and Optimal Land Use

In agriculture, achieving optimum land utilization is a critical factor for sustainable food production. Farmers must consider factors such as soil quality, climate, and topography to determine the best crops to cultivate on a particular piece of land. Precision agriculture, which involves using technology such as GPS-guided tractors and sensors, has emerged as a tool to optimize land use by allowing farmers to tailor inputs like water and fertilizers to specific areas of a field.

Case Study: Precision Agriculture in the United States

Precision agriculture has gained widespread adoption in the United States, where farmers use advanced technologies to optimize the use of land, water, and other resources. GPS technology allows farmers to precisely plant seeds, apply fertilizers, and manage irrigation, resulting in increased efficiency and reduced environmental impact. By tailoring inputs to specific areas of a field based on soil characteristics and crop needs, farmers can achieve higher yields with lower resource use.

4.4.2 Manufacturing and the Law of Increasing Returns

The Law of Increasing Returns is particularly relevant in manufacturing, where specialization and division of labor can lead to significant efficiency gains. One notable example is the automotive industry, where assembly lines and specialized tasks contribute to higher production rates. As more workers specialize in specific tasks, the overall productivity of the manufacturing process increases, leading to cost savings and economies of scale.

Case Study: Toyota Production System

The Toyota Production System (TPS) is a renowned example of the Law of Increasing Returns in action. TPS emphasizes lean manufacturing principles, including just-in-time production and continuous improvement. By optimizing the assembly line and reducing waste, Toyota has been able to achieve higher production efficiency and quality. The principles of TPS have been widely adopted in the manufacturing industry globally, demonstrating the effectiveness of increasing returns in enhancing productivity.

4.4.3 Technology and Optimal Capital Utilization

The rapid pace of technological advancement has a profound impact on the concept of optimum capital utilization. Businesses must continually assess and adopt new technologies to remain competitive and achieve optimal efficiency in production. The shift toward Industry 4.0, characterized by the integration of digital technologies and automation in manufacturing processes, exemplifies the ongoing evolution in capital utilization.

Case Study: Industry 4.0 Implementation in Germany

Germany, known for its strong manufacturing sector, has embraced Industry 4.0 to enhance capital utilization. The adoption of smart manufacturing technologies, such as the Internet of Things (IoT), artificial intelligence, and robotics, has allowed German manufacturers to optimize production processes. Real-time data analytics and predictive maintenance contribute to minimizing downtime and maximizing the lifespan of capital assets. This case study illustrates how leveraging technology can lead to optimal capital utilization and sustained competitiveness.

4.5 Challenges and Considerations

The application of economic theories, including the Laws of Return, is not without its challenges and considerations. One significant challenge lies in the dynamic and interconnected nature of economic systems, which can introduce complexities that theoretical models may not fully capture. External factors such as changes in technology, market conditions, and government policies can influence the practical application of economic principles. Additionally, the assumption of ceteris paribus (all other things being equal) in economic models may not always hold in the real world. Adapting economic theories to diverse contexts and recognizing the uniqueness of individual industries or regions poses another challenge. Moreover, ethical considerations and social impacts must be taken into account, especially when making decisions that affect communities and the environment. Striking a balance between economic efficiency and broader societal well-being is a critical consideration. These challenges underscore the need for a nuanced and context-specific approach to economic decision-making, acknowledging the limitations of theoretical frameworks and addressing the multifaceted dynamics of real-world scenarios.

4.5.1 Environmental Sustainability

As the global community grapples with environmental challenges, achieving optimum land utilization must align with principles of sustainability. Uncontrolled exploitation of natural resources can lead to deforestation, soil degradation, and loss of biodiversity. Striking a balance between economic development and environmental conservation is essential to ensure that future generations have access to the resources necessary for their well-being.

4.5.2 Labor Market Dynamics

Optimizing labor utilization requires addressing the challenges posed by evolving labor market dynamics. The rise of automation and artificial intelligence raises concerns about job displacement and the need for reskilling the workforce. Policymakers and businesses must collaborate to create frameworks that support a smooth transition for workers into new roles and industries, ensuring that the benefits of technological advancement are shared equitably.

4.5.3 Technological Disparities

While advanced technologies contribute to optimal capital utilization, there is a risk of creating technological disparities between developed and developing regions. Access to cutting-edge technologies may be limited in certain areas, hindering the ability of businesses to achieve optimum efficiency. Efforts to bridge the technological gap and promote inclusive development are essential for ensuring that the benefits of technological progress are shared globally.

4.6 Future Trends and Opportunities

The future of economic theory, including the application of principles like the Laws of Return, holds promising trends and opportunities. As technology continues to advance, there is an increasing emphasis on leveraging data analytics, artificial intelligence, and machine learning to refine economic models and predictions. This shift towards more sophisticated modeling allows for a better understanding of complex interactions within economic systems. Additionally, the growing awareness of environmental sustainability is influencing economic paradigms, fostering the development of models that account for ecological impacts and support green practices. The rise of the digital economy and the globalization of markets present opportunities to explore new dimensions of economic theories, particularly in understanding cross-border dynamics and the implications of technological advancements on international trade. As economies evolve, there is a greater recognition of the importance of inclusive growth and social equity, prompting the integration of social and ethical considerations into economic models. Future opportunities lie in the interdisciplinary collaboration between economists, technologists, and environmental scientists to develop holistic frameworks that address the multifaceted challenges of the 21st century.

4.6.1 Sustainable Agriculture

The future of agriculture lies in sustainable practices that optimize land use while preserving the environment. Precision agriculture will continue to evolve, incorporating advancements in sensors, artificial intelligence, and data analytics. Techniques such as vertical farming and agroforestry present innovative opportunities to maximize food production while minimizing the ecological footprint.

4.6.2 Digital Transformation in Manufacturing

The ongoing digital transformation in manufacturing, driven by technologies like the Industrial Internet of Things (IIoT) and machine learning, holds immense potential for optimizing capital utilization. Smart factories with interconnected systems will enable real-time monitoring, predictive maintenance, and efficient resource allocation. The adoption of digital twins, virtual representations of physical assets, will further enhance decision-making and production efficiency.

4.6.3 Entrepreneurship and Innovation Ecosystems

The cultivation of robust entrepreneurship and innovation ecosystems is critical for fostering economic growth. Governments, educational institutions, and businesses will play pivotal roles in creating environments that nurture entrepreneurial talent. Initiatives such as startup incubators, access to funding, and supportive regulatory frameworks will continue to shape the landscape for optimum entrepreneurship.

4.7 Conclusion

The concept of factors of production, encompassing land, labor, capital, and entrepreneurship, forms the foundation of economic activities. Achieving optimum utilization of these factors is essential for maximizing efficiency and output in the production process. The laws of return—increasing returns, constant returns, and diminishing returns—provide insights into the dynamics of input-output relationships, guiding decision-makers in resource allocation and production strategies.

As we navigate a rapidly evolving global landscape, addressing challenges such as environmental sustainability, labor market dynamics, and technological disparities becomes imperative. Embracing future trends and opportunities, including sustainable agriculture, digital transformation in manufacturing, and the development of robust entrepreneurship ecosystems, will shape the trajectory of economic development. In essence, the pursuit of optimum in factors of production is not a static endeavor but a dynamic process that requires continuous adaptation to changing circumstances. By understanding and harnessing the principles of optimum and the laws of return, individuals, businesses, and societies can strive for economic prosperity while mindful of the broader impact on the planet and its inhabitants.

Suggested Questions with Solutions

1. Question: What are the primary factors of production, and how do they contribute to the production process?

Solution: The primary factors of production are land, labor, capital, and entrepreneurship. Land provides natural resources, labor contributes human effort, capital includes physical and financial assets, and entrepreneurship combines these factors to create goods and services.

2. Question: Explain the concept of optimum utilization of factors of production. How can businesses achieve efficiency in resource allocation?

Solution: Optimum utilization refers to the efficient allocation of resources to maximize output. Businesses can achieve efficiency by employing the right combination of factors, adopting advanced technology, minimizing waste, and optimizing production processes.

3. Question: Discuss the importance of technological advancements in influencing the efficiency of factors of production.

Solution: Technological advancements play a crucial role in enhancing the efficiency of factors of production. Innovations in technology can lead to increased productivity, reduced costs, and improved utilization of resources.

4. Question: What are the laws of returns to scale, and how do they impact the production process?

Solution: The laws of returns to scale include increasing returns, constant returns, and diminishing returns. Increasing returns occur when all inputs are increased, leading to a more than proportionate increase in output. Constant returns maintain a proportional relationship, while diminishing returns occur when additional inputs result in proportionately smaller output increases.

5. Question: How does specialization contribute to the efficiency of factors of production?

Solution: Specialization involves focusing on specific tasks or functions. It enhances efficiency by allowing individuals or businesses to develop expertise, leading to increased productivity and improved allocation of resources.

6. Question: Explore the role of human capital in optimizing factors of production.

Solution: Human capital, which includes knowledge, skills, and education, plays a vital role in optimizing factors of production. Well-educated and skilled workers contribute to higher productivity and innovation, positively impacting the overall production process.

7. Question: Discuss the challenges businesses may face in achieving optimum utilization of factors of production in a globalized economy.

Solution: In a globalized economy, businesses may face challenges such as increased competition, varying labor costs, and complex supply chains. Overcoming these challenges requires strategic planning, adaptability, and effective resource management.

8. Question: How can government policies influence the optimal use of factors of production and address market failures in the economy?

Solution: Government policies, such as tax incentives, subsidies, and regulations, can influence the optimal use of factors of production. These policies can address market failures, promote fair competition, and create an environment conducive to efficient resource allocation.

9. Question: Assume a box manufacturer is producing in the short run when equipment is fixed. **The manufacturer recognizes that as the number of laborers used in the production process increases from 1 to 8, the number of chairs produced changes as follows: 14, 27, 38, 46, 52, 57, 61, 64. Calculate the marginal and average product of labor for this production function. Does this production function exhibit diminishing returns to labor?** **Explain with the possible reason.**

Solutions: The average product of labor (APL) is equal to $\frac{Q}{L}$. The marginal product of labor (MPL) is equal to $\frac{\Delta Q}{\Delta L}$, the change in output divided by the change in labor input. For this production process we have:

L	Q	APL	MPL
0	0	-	-
1	14	14	14
2	27	13.50	13
3	38	12.67	11
4	46	11.50	8
5	52	10.40	6
6	57	9.50	5
7	61	8.71	4
8	64	8	3

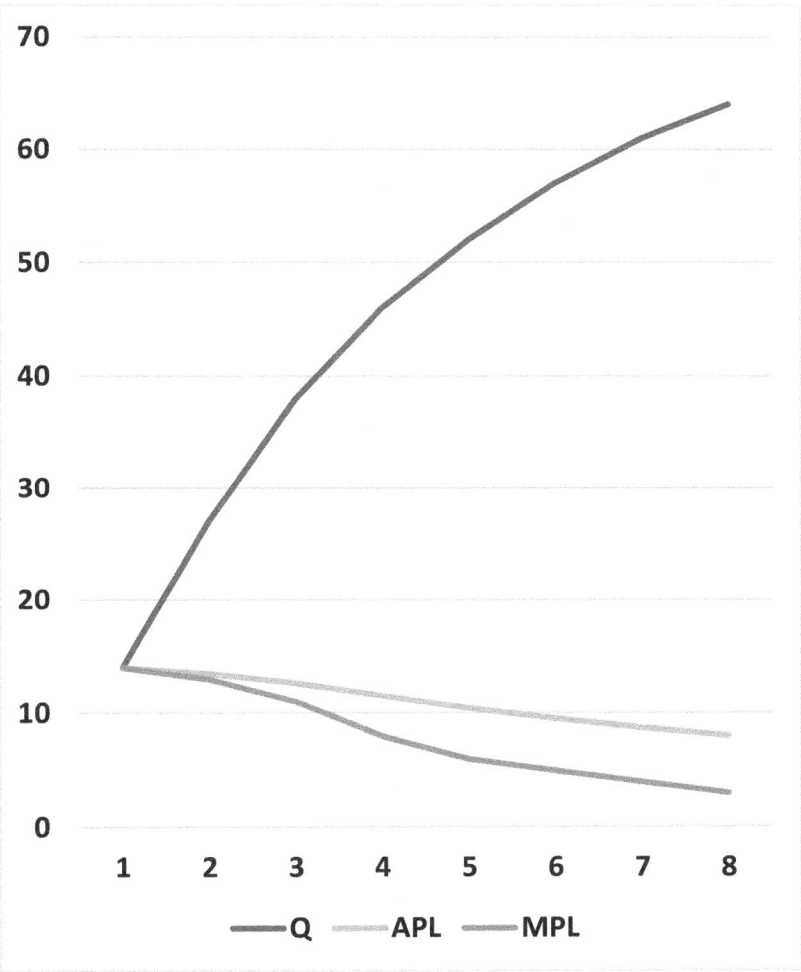

Fig. 1: Plot of Production Level, Average Product of Labor, and Marginal Product of Labor

This production process exhibits diminishing returns to labor. The marginal product of labor, the extra output produced by each additional worker, diminishes as workers are added. Labor's diminishing marginal product may arise from congestion in the box manufacturer's factory. Since more laborers are using the same, fixed amount of capital, it is possible that they could get in each other's way, decreasing efficiency and the amount of output.

10. Question: Fill in the gaps in the table below.

Quantity of Variable Input	Total Output	Marginal Product of Variable Input	Average Product of Variable Input
0	0	-	-
1	400		
2			250
3		130	
4	800		
5		150	
6			200
7	1470		
8		290	
9		40	
10			210

Solution:

Quantity of Variable Input	Total Output	Marginal Product of Variable Input	Average Product of Variable Input
0	0	-	-
1	400	400	400
2	500	100	250
3	630	130	210
4	800	170	200
5	950	150	190
6	1200	250	200
7	1470	270	210
8	1760	290	220
9	1800	40	200
10	2100	300	210

11. Question: The marginal product of labor in the production of buckets is 30 buckets per hour. The marginal rate of technical substitution of hours of labor for hours of machine-capital is 0.3. What is the marginal product of capital?

Solution: The marginal rate of technical substitution computed as:

$$Marginal\ Rate\ of\ Technical\ Substitution\ (MRTS)$$

$$= \frac{Marginal\ Product\ of\ Labor\ (MPL)}{Marginal\ Product\ of\ Capital\ (MPC)}$$

Given, MRTS = 0.3,
MPL = 30.

So, MPC $= \frac{30}{0.3} = 100$.

12. Question: Write briefly about the Cobb-Douglas production function. The production function for the gears of MECHANICA, Inc., is given by $Q = 20K^{0.5}L^{0.5}$, where Q is the number of laptops produced per day, K is hours of machine time, and L is hours of labor input. MECHANICA's competitor, AUTOMATE, Inc., is using the production function $Q = 20K^{0.7}L^{0.3}$.
A) If both companies use the same amounts of capital and labor, which will generate more output?
B) Assume that capital is limited to 8 machine hours but labor is unlimited in supply. In which company is the marginal product of labor greater? Explain.

Solution: The Cobb-Douglas production function is a widely used economic model that describes the relationship between the inputs (factors of production) and the output of a production process. Named after economists Paul Douglas and Charles Cobb, this function is commonly employed in the field of macroeconomics and microeconomics to analyze production and growth.

The Cobb-Douglas production function takes the following general form:

$$Q = A \cdot L^\alpha \cdot K^\beta$$

Here:
- Q represents the output or production.
- A is the total factor productivity, representing the technological level and efficiency of production.
- L denotes the quantity of labor input.
- K represents the quantity of capital input.
- α and β are positive constants that represent the output elasticities of labor and capital, respectively. These parameters indicate the percentage change in output resulting from a 1% change in the respective input.

The Cobb-Douglas function has several key characteristics:

1. Constant Returns to Scale: If $\alpha + \beta = 1$, the production function exhibits constant returns to scale. This means that if both labor and capital are increased by a certain percentage, output will increase by the same percentage.

2. Increasing Returns to Scale: If $\alpha + \beta > 1$, the production function demonstrates increasing returns to scale. In this case, a percentage increase in inputs leads to a more than proportionate increase in output.

3. Decreasing Returns to Scale: If $\alpha + \beta < 1$, the production function shows decreasing returns to scale. Here, a percentage increase in inputs results in a less than proportionate increase in output.

The Cobb-Douglas production function is valuable for understanding the relationship between inputs and output in various economic activities. It provides a simplified yet powerful tool for economists to analyze and model production processes, growth, and productivity in both individual firms and entire economies.

A) Let Q_1 be the output of MECHANICA, Inc., Q_2, be the output of AUTOMATE, Inc., and X be the same equal amounts of capital and labor for the two firms. Then, according to their production functions,

$$Q_1 = 20\ X^{0.5}X^{0.5} = 20\ X$$

$$Q_2 = 20\ X^{0.7}X^{0.3} = 20\ X$$

Because $Q_1 = Q_2$, both firms generate the same output with the same inputs. Note that if the two firms both used the same amount of capital and the same amount of labor, but the amount of capital was not equal to the amount of labor, then the two firms would not produce the same level of output. In fact, if K>L then Q_2>Q_1.

B) With capital limited to 8 machine units, the production functions become $Q_1 = 56.57L^{0.5}$ and $Q_2 = 85.74L^{0.3}$. To determine the production function with the highest marginal productivity of labor, consider the following table:

L	Q_1	MPL_1	Q_2	MPL_2
0	0	-	0	-
1	56.57	56.57	85.74	85.74
2	80	23.43	105.56	19.82
3	97.98	17.98	119.21	13.65
4	113.14	15.16	129.96	10.75

For each unit of labor above 1, the marginal productivity of labor is greater for the first firm, MECHANICA, Inc.

5 UNDERSTANDING DEMAND AND ITS DYNAMICS

5.1 Demand

In the realm of economics and business, demand represents the quantity of a good or service that consumers are willing and able to purchase at different price levels within a specific time frame. It is a fundamental concept that plays a pivotal role in shaping market dynamics. The study of demand is essential for businesses as it serves as the foundation for various strategic decisions, including pricing, production planning, and marketing strategies.

Understanding demand involves recognizing the factors that influence consumers' purchasing decisions. These factors can include the price of the product, the income of consumers, their preferences, and external economic conditions. The demand curve, a graphical representation of the relationship between price and quantity demanded, is a crucial tool in visualizing and analyzing these dynamics.

Demand is a fundamental economic concept that plays a crucial role in shaping market dynamics and influencing business strategies. In the realm of economics, demand refers to the quantity of a good or service that consumers are willing and able to purchase at a given price and within a specific time frame. Understanding the intricacies of demand is essential for businesses, policymakers, and economists, as it serves as a cornerstone for decision-making, market analysis, and economic forecasting.

The law of demand is a foundational principle that governs the relationship between price and quantity demanded. According to this law, all else being equal, as the price of a good or service decreases, the quantity demanded increases, and conversely, as the price rises, the quantity demanded decreases. This inverse relationship is depicted on a demand curve, where the x-axis represents the quantity demanded, and the y-axis represents the price. The demand curve typically slopes downward from left to right, illustrating the law of demand.

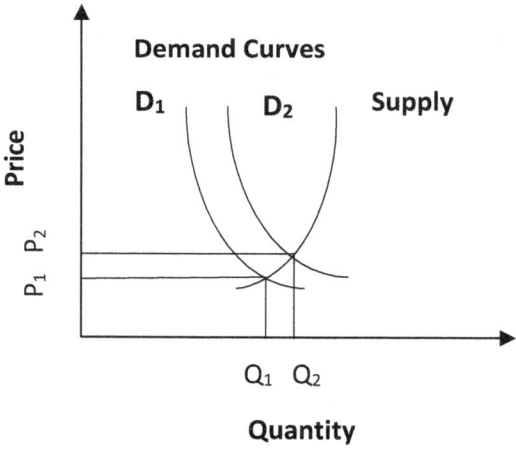

Fig. 2: Demand Curve

Several factors influence demand, and they can be broadly categorized into price-related factors and non-price-related factors. Price elasticity of demand measures how sensitive the quantity demanded is to changes in price. If a small change in price leads to a proportionally larger change in quantity demanded, the good is considered elastic. In contrast, if a change in price has a relatively smaller impact on quantity demanded, the good is considered inelastic. Elasticity is an important concept for businesses when setting prices, as it helps predict consumer responsiveness to price changes.

Non-price factors that affect demand include consumer income, preferences and tastes, the prices of related goods (substitutes and complements), and external factors such as demographics and cultural influences. For example, if consumers' incomes rise, they may be more willing to purchase higher-priced goods, leading to an increase in demand. Similarly, changes in consumer preferences, fashion trends, or cultural shifts can significantly impact demand for certain products.

The concept of elasticity also extends to income elasticity of demand, which measures the responsiveness of quantity demanded to changes in consumer income. Goods are classified as normal or inferior based on their income elasticity. Normal goods experience an increase in demand as consumer incomes rise, while inferior goods see a decrease in demand as incomes increase.

Understanding cross-price elasticity is vital for analyzing the relationships between different goods. Substitutes are goods that can be used in place of each other, and an increase in the price of one substitute typically leads to an increase in demand for the other. Complements, on the other hand, are goods that are consumed together, and an increase in the price of one complement often results in a decrease in demand for the other.

Apart from these factors, the concept of demand also extends beyond individual consumers to aggregate demand, which represents the total quantity of goods and services demanded in an economy at a given price level and in a specific time period. Aggregate demand is influenced by factors such as overall economic activity, interest rates, government spending, and international trade. Policymakers use insights into aggregate demand to formulate economic policies that aim to stabilize and stimulate economic growth.

Demand is not a static concept; it evolves over time in response to changing economic conditions, technological advancements, and shifts in consumer behavior. In recent years, the advent of e-commerce and digital platforms has transformed the way consumers interact with products and services, influencing the demand patterns across various industries. The ability to adapt to these changes and accurately anticipate shifts in demand is critical for businesses seeking sustained success in dynamic markets.

The COVID-19 pandemic, for instance, had profound effects on demand across multiple sectors. Lockdowns, supply chain disruptions, and changes in consumer behavior led to fluctuations in demand for certain products and services. The demand for online collaboration tools surged as remote work became the norm, while sectors such as travel and hospitality experienced a sharp decline in demand.

In the context of business strategy, understanding and effectively managing demand is a key determinant of success. Companies employ various strategies to stimulate and capture demand, such as dynamic pricing, bundling, and promotional activities. Dynamic pricing, often used in industries like airline travel and hospitality, involves adjusting prices in real-time based on factors such as demand levels, competitor pricing, and other market conditions.

Product bundling is another strategy to influence demand, where related products or services are offered together at a discounted price. This approach can encourage consumers to purchase a bundle rather than individual items, thereby boosting overall demand. Promotional activities, including advertising and sales promotions, are employed to create awareness and generate interest in a product or service, ultimately driving demand.

The field of marketing revolves around creating and fulfilling demand. Through market research, companies gain insights into consumer needs and preferences, enabling them to tailor their products and messaging to align with demand. Successful marketing campaigns not only create awareness but also shape consumer perceptions and influence purchasing decisions.

The concept of demand also extends to the labor market, where demand for specific skills and expertise drives employment trends. In a rapidly evolving technological landscape, the demand for workers with digital skills, data analytics, and programming capabilities has surged. Governments, educational institutions, and businesses are increasingly focused on addressing this demand through initiatives such as workforce training programs and educational curriculum enhancements.

From a global perspective, international trade is influenced by variations in demand for goods and services between countries. Comparative advantage, a concept introduced by economist David Ricardo, suggests that countries should specialize in producing goods for which they have a relative efficiency, leading to increased efficiency and overall gains in global welfare. Understanding international demand patterns is crucial for countries to optimize their production and trade strategies.

In conclusion, the concept of demand is a multifaceted and dynamic force that shapes economic activities at various levels. It is a central element in economic theory, influencing pricing strategies, market dynamics, and policymaking. Businesses that grasp the nuances of demand and respond adeptly to changing market conditions are better positioned for success in today's fast-paced and interconnected global economy. As technology continues to evolve and consumer behaviors shift, the ability to analyze and anticipate demand patterns will remain a key determinant of competitiveness and sustainability for businesses and economies alike.

5.2 Elasticity of Demand

Elasticity of demand measures how responsive the quantity demanded is to changes in price. Mathematically, it is expressed as the percentage change in quantity demanded divided by the percentage change in price. The concept of elasticity categorizes demand into elastic, inelastic, or unitary elastic.

Elastic Demand: When the percentage change in quantity demanded is greater than the percentage change in price, demand is considered elastic. In this scenario, consumers are highly responsive to price changes. Luxury goods often exhibit elastic demand, as consumers may reduce their purchases significantly when prices increase.

Inelastic Demand: In contrast, inelastic demand occurs when the percentage change in quantity demanded is less than the percentage change in price. This suggests that consumers are relatively insensitive to price changes. Necessities like medications or certain utilities often demonstrate inelastic demand.

Unitary Elasticity: Unitary elasticity exists when the percentage change in quantity demanded is equal to the percentage change in price. This implies that total spending by consumers remains constant despite price fluctuations.

Understanding elasticity is crucial for businesses when setting prices. If demand is elastic, a price decrease could lead to a significant increase in quantity demanded, potentially resulting in higher overall revenue. On the other hand, in markets with inelastic demand, price increases might not significantly reduce the quantity demanded, providing an opportunity for increased revenue.

1. Price Elasticity of Demand (PED):

$$PED = \frac{\% \text{ Change in quantity demanded}}{\% \text{ Change in price}}$$

= (Change in quantity demanded / Initial quantity demanded)/(Change in price / Initial price)

2. Income Elasticity of Demand (YED):

$$YED = \frac{\% \text{ Change in quantity demanded}}{\% \text{ Change in income}}$$

= (Change in quantity demanded / Initial quantity demanded)/(Change in income / Initial income)

3. Cross-Price Elasticity of Demand (XED):

$$\text{XED} = \frac{\%\text{ Change in quantity demanded of good A}}{\%\text{ Change in price of good B}}$$

= (Change in quantity demanded of good A/Initial quantity demanded of good A)/(Change in Price of good B / Initial Price of good B)

5.3 Demand Estimation

Demand estimation involves predicting the quantity of a good or service that consumers will demand at various price points. Accurate demand estimation is vital for businesses to make informed decisions regarding production levels, inventory management, and pricing strategies. Several methods are employed in demand estimation:

1. Surveys: Gathering direct feedback from potential consumers through surveys is a common method. Surveys can provide insights into consumer preferences, purchasing behavior, and the factors influencing their decisions. Let's consider a practical example of a survey for demand estimation, specifically for a new mobile app that aims to provide fitness and health tracking services. The goal of the survey is to understand potential users' preferences, willingness to pay, and the factors influencing their decision to use such an app.

Survey Title: HealthTracker App User Preferences and Demand Estimation

Introduction:
"Thank you for participating in our survey! Your input is valuable as we aim to develop a new mobile app, HealthTracker, to help individuals monitor and improve their fitness and overall health. Please take a few minutes to share your thoughts and preferences with us."

Section 1: Demographics
1.1. Age: [Open-ended response]

1.2. Gender:
- Male

- Female
- Other

1.3. Current Fitness Level:
- Sedentary
- Lightly active
- Moderately active
- Very active

Section 2: Mobile App Usage

2.1. How often do you use health and fitness apps?
- Daily
- Weekly
- Monthly
- Rarely
- Never

2.2. What features do you currently find most valuable in health and fitness apps? (Select all that apply)
- Activity tracking
- Nutrition tracking
- Workout routines
- Social community
- Goal setting and tracking
- Challenges and rewards

Section 3: HealthTracker App Concept

3.1. Have you ever used a health and fitness tracking app before?
- Yes
- No

3.2. What features would you like to see in a health and fitness app like HealthTracker? [Open-ended response]

3.3. On a scale of 1 to 10, how likely are you to use HealthTracker when it becomes available? (1 being not likely at all, 10 being extremely likely)

Section 4: Pricing and Monetization

4.1. How much would you be willing to pay for a monthly subscription to HealthTracker?
- $0-$5
- $6-$10
- $11-$15
- $16-$20
- $20+

4.2. What factors would influence your decision to subscribe to HealthTracker? (Select all that apply)
- Affordable pricing
- Free trial period
- Exclusive premium features
- Positive reviews and ratings
- Referral discounts

Section 5: Closing Questions

5.1. Do you have any additional comments or suggestions regarding the HealthTracker app concept? [Open-ended response]

5.2. Would you be interested in participating in beta testing for HealthTracker?

Thank you for taking the time to complete our survey! Your feedback is essential in helping us create a health and fitness app that meets the needs and preferences of our users.

2. Focus Groups: These involve small groups of individuals discussing their perceptions and preferences regarding a product or service. Focus groups can offer qualitative insights that may not be apparent through quantitative methods alone.

3. Analysis of Sales Data: Historical sales data provides valuable information about past consumer behavior. By analyzing trends and patterns, businesses can make informed predictions about future demand. Demand forecasting involves predicting future demand for a product or service. Various methods can be used for demand forecasting, and the choice of method depends on the nature of the data and the industry. Here are some common demand forecasting formulae:

1. **Simple Moving Average:**

$$Forecast = \frac{1}{n} \sum_{i=1}^{n} Demand_{t-i}$$

Where, n is the number of periods.

2. Weighted Moving Average:

$$Forecast = w_1 \times Demand_{t-1} + w_2 \times Demand_{t-2} + w_3 \times Demand_{t-3} + \cdots + w_n \times Demand_{t-n}$$

Where, $w_1, w_2, w_3 \ldots w_n$ are weights assigned to each period.

3. Exponential Smoothing:

$$Forecast_t = \alpha \times Demand_t + (1 - \alpha) \times Forecast_{t-1}$$

Where, $0 \leq \alpha \leq 1$ is the smoothing constant.

4. Trend Projection (Linear Regression):

$$Forecast = a + b \times T$$

Where, a is the intercept, b is the slope, and T represents the time periods.

5. Seasonal Adjustment:

$$\text{Seasonal Factor} = \frac{\text{Average Demand in Season}}{\text{Average Demand over All Periods}}$$

Adjust the demand for each period by multiplying it by the corresponding seasonal factor.

6. Causal Models:

For models that consider multiple variables influencing demand (e.g., price, advertising, etc.), a causal model can be used. An example might be:

$$Q = \beta_0 + \beta_1 P + \beta_2 I + \beta_3 A + \beta_4 T + \varepsilon$$

Where, $\beta_0, \beta_1, \beta_2, \beta_3, \beta_4$ are coefficients and ε is the error term.

- Q be the quantity demanded.
- P be the price of the product.
- I be the income of consumers.
- A be advertising expenditure.
- T be the time period.

These formulas provide a glimpse into different methods for demand forecasting. The appropriate method depends on the characteristics of the data available and the specific requirements of the forecasting task. It's common for businesses to use a combination of these methods or more advanced forecasting techniques, such as machine learning models, depending on the complexity of the forecasting problem.

4. Statistical Techniques: Regression analysis and other statistical methods can be employed to model the relationship between price and quantity demanded. These techniques help in quantifying the impact of various factors on demand.

By integrating these methods, businesses can develop a more comprehensive understanding of consumer behavior and market trends, enabling them to respond proactively to changes in demand.

5.4 Market Research

Market research is a multifaceted process that involves systematically gathering, analyzing, and interpreting information about a market. The goal is to gain insights into consumer needs, preferences, and behaviors, as well as to understand the competitive landscape. Effective market research is instrumental in developing successful marketing strategies, launching new products, and staying ahead of industry trends.

5.4.1 Key Components of Market Research

1. Consumer Behavior Analysis: Understanding how consumers make purchasing decisions is at the core of market research. This includes studying factors such as motivation, perception, and attitudes that influence buying choices.

2. Competitor Analysis: Assessing the strengths and weaknesses of competitors helps businesses identify opportunities and threats in the market. This analysis includes studying competitors' products, pricing strategies, and market positioning.

3. Trend Analysis: Identifying current and emerging trends is crucial for staying relevant in a dynamic market. This involves monitoring changes in consumer preferences, technological advancements, and societal shifts.

4. SWOT Analysis: Conducting a SWOT (Strengths, Weaknesses, Opportunities, Threats) analysis provides a holistic view of a business and its position in the market. This analysis aids in strategic decision-making by highlighting internal and external factors.

5. Market Segmentation: Dividing the market into segments based on demographics, psychographics, or behavior helps businesses tailor their products and marketing strategies to specific consumer groups.

6. Feasibility Studies: Before launching a new product or entering a new market, businesses conduct feasibility studies to assess the viability of their plans. This includes analyzing potential risks, resource requirements, and market acceptance.

Market research is an ongoing process, and businesses that invest in continuous analysis and adaptation are better positioned to navigate the complexities of the market successfully.

5.5 The Dynamics of Supply and Industrial Costs

The dynamics of supply and industrial costs play a pivotal role in shaping economic landscapes. Supply, the quantity of goods or services available in the market, is influenced by factors such as production capabilities, technology, and resource availability. As the law of supply suggests, a direct relationship exists between the price of a product and the quantity producers are willing to supply – higher prices often lead to increased supply. Industrial costs, on the other hand, encompass the expenses incurred in the production process, including raw materials, labor, and overhead. These costs fluctuate based on market conditions, technological advancements, and regulatory changes. Understanding the interplay between supply and industrial costs is crucial for businesses to set competitive prices, optimize production processes, and navigate market dynamics. Efficient supply chain management, innovations in production technologies, and strategic cost-cutting measures contribute to maintaining competitiveness in industries. The continuous evaluation and adaptation to these dynamics are essential for businesses to thrive in ever-changing economic environments.

5.5.1 Supply

Supply, in economic terms, refers to the quantity of a good or service that producers are willing and able to offer for sale at different price levels. The supply curve, like the demand curve, illustrates the relationship between price and quantity supplied. The law of supply states that, all else being equal, an increase in price will lead to an increase in quantity supplied, while a decrease in price will result in a decrease in quantity supplied. Several factors influencing supply are:

1. Production Costs: The cost of producing goods or services is a significant determinant of supply. If production costs increase, businesses may reduce the quantity supplied unless they can pass the increased costs on to consumers through higher prices.

2. Technology: Advances in technology can enhance production efficiency, leading to increased supply. Automation, improved machinery, and streamlined processes can positively impact a firm's ability to produce more with the same resources.

3. Input Prices: Prices of raw materials and other inputs directly affect production costs. Fluctuations in these input prices can influence the profitability of producing goods and, consequently, the quantity supplied.

4. Government Policies: Regulations and policies imposed by governments can have a substantial impact on supply. For example, environmental regulations might increase production costs, affecting the supply of certain goods.

5. Expectations of Future Prices: Producers' expectations about future prices can influence their current supply decisions. If producers anticipate that prices will rise in the future, they may reduce current supply to capitalize on higher future profits.

Understanding supply dynamics is crucial for maintaining market equilibrium. When demand and supply are in balance, the market efficiently allocates resources, and prices stabilize. However, imbalances can lead to surpluses or shortages, necessitating adjustments in production or pricing strategies.

5.5.2 Industrial Costs

Industrial costs encompass the various expenses incurred in the production of goods or services. Managing these costs efficiently is essential for businesses to remain competitive and sustainable. Industrial costs can be broadly categorized into the following:

1. Raw Materials: The cost of acquiring raw materials is a significant component of industrial costs. Fluctuations in commodity prices, geopolitical events, and supply chain disruptions can impact the availability and cost of raw materials.

2. Labor Costs: The expenses associated with hiring and compensating labor constitute a substantial portion of industrial costs. This includes wages, benefits, and training expenses.

3. Overhead Costs: Overhead costs encompass indirect expenses such as rent, utilities, insurance, and administrative expenses. Managing overhead efficiently is crucial for maintaining overall cost-effectiveness.

4. Technology and Equipment: Investments in technology and machinery can enhance production efficiency but also contribute to industrial costs. The initial capital outlay and ongoing maintenance expenses must be carefully managed to ensure a positive return on investment.

5. Transportation and Distribution Costs: Getting finished products to market involves transportation and distribution expenses. These costs are influenced by factors such as fuel prices, infrastructure, and the complexity of the supply chain.

6. Regulatory Compliance: Adhering to regulatory requirements can add compliance costs to industrial operations. These may include costs associated with environmental regulations, safety standards, and quality control measures.

Efficient management of industrial costs involves finding the right balance between cost reduction and maintaining product quality. Businesses often employ cost-control measures, such as process optimization, bulk purchasing, and strategic sourcing, to enhance efficiency and competitiveness.

5.5.3 The Interplay of Demand, Supply, and Industrial Costs

Understanding the interplay between demand, supply, and industrial costs is crucial for businesses to formulate effective strategies and make informed decisions. Here's how these elements interact:

1. Market Equilibrium: The equilibrium in the market is achieved when the quantity demanded equals the quantity supplied at a specific price. This equilibrium price reflects the point where consumers are willing to pay, and producers are willing to sell, resulting in a stable market.

2. Price Adjustments: Changes in demand and supply can lead to shifts in the equilibrium price. For instance, if demand increases, the price tends to rise unless the supply can also increase to meet the heightened demand. Conversely, if supply decreases, prices may increase unless demand decreases or alternative supply sources are found.

3. Elasticity's Impact on Pricing: Elasticity of demand influences pricing strategies. In markets with elastic demand, businesses might choose lower prices to stimulate demand and increase overall revenue. In markets with inelastic demand, businesses might have more flexibility to raise prices without significantly affecting quantity demanded.

4. Cost-Volume-Profit Analysis: Understanding the relationship between production costs, volume, and profit is essential. Cost-volume-profit (CVP) analysis helps businesses assess the impact of changes in production levels, costs, and pricing on overall profitability.

5. Strategic Decision-Making: Businesses use insights from demand estimation and market research to make strategic decisions about production levels and market positioning. If demand for a product is projected to increase, a business may invest in expanding production capacity or introducing variations of the product to meet diverse consumer needs.

6. Supply Chain Management: Efficient supply chain management is critical for minimizing industrial costs. Streamlining processes, negotiating favorable terms with suppliers, and optimizing logistics contribute to cost reduction and increased competitiveness.

7. Risk Management: External factors, such as economic downturns, natural disasters, or geopolitical events, can impact demand, supply, and industrial costs. Businesses need robust risk management strategies to navigate uncertainties and adapt to changing market conditions.

In summary, the dynamics of demand, supply, and industrial costs are interconnected and form the backbone of a market-oriented economy. Businesses that can effectively analyze and respond to these dynamics are better equipped to make strategic decisions, stay competitive, and achieve long-term success. The integration of these concepts provides a holistic understanding of market behavior, enabling businesses to adapt and thrive in dynamic and ever-changing environments.

Suggested Questions with Solutions

1. Question: Explain the concept of price elasticity of demand. How is it calculated, and what does a value greater than 1, less than 1, and equal to 1 indicate?

Solution: Price elasticity of demand measures the responsiveness of quantity demanded to a change in price. The formula is:

$$\text{Price Elasticity of Demand (PED)} = \frac{\%\text{change in quantity demanded}}{\%\text{change in price}}$$

- If $|PED| > 1$, demand is elastic (responsive to price changes).
- If $|PED| < 1$, demand is inelastic (less responsive to price changes).
- If $|PED| = 1$, demand is unit elastic.

2. Question: What factors can influence the demand for a product? Provide examples and explain how they affect demand.

Solution: Factors influencing demand include:
- Price of the product.
- Income levels.
- Prices of related goods (substitutes and complements).
- Consumer preferences and tastes.
- Advertising and marketing.
- Seasonal trends.
- Government policies.

3. Question: Describe the concept of cross-price elasticity of demand. Provide an example of substitute goods and another example of complementary goods.

Solution: Cross-price elasticity of demand (XED) measures how the quantity demanded of one good changes in response to a change in the price of another good. It is calculated as:

$$XED = \frac{\%\text{change in quantity demanded of good A}}{\%\text{change in price of good B}}$$

- Substitute goods (positive XED): If the price of coffee increases, the demand for tea may increase.
- Complementary goods (negative XED): If the price of smartphones increases, the demand for smartphone cases may decrease.

4. Question: How does income elasticity of demand help classify goods as normal or inferior? Provide examples.

Solution: Income elasticity of demand (YED) measures how the quantity demanded changes in response to a change in income. Classification:
- YED > 0 for normal goods (e.g., luxury items like sports cars).
- YED < 0 for inferior goods (e.g., generic products like instant noodles).

5. Question: Discuss the role of expectations in influencing consumer demand. How do expectations impact purchasing decisions?

Solution: Expectations about future prices, income levels, or product availability can influence current demand. For example, if consumers expect prices to rise in the future, they may increase their current demand to avoid higher costs later.

6. Question: Explain the concept of elasticity of supply. How is it different from price elasticity of demand?

Solution: Elasticity of supply measures how the quantity supplied of a good changes in response to a change in price. It is calculated as:

$$\text{Elasticity of Supply} = \frac{\%\text{change in quantity supplied}}{\%\text{change in price}}$$

The key difference is that price elasticity of demand focuses on the buyer's perspective, while elasticity of supply looks at the seller's response to price changes.

7. Question: Describe the impact of advertising on demand. How can effective advertising strategies influence consumer behavior?

Solution: Advertising can increase awareness, create brand loyalty, and influence consumer preferences. Effective advertising can lead to higher demand for a product by emphasizing its unique features, benefits, or value proposition.

8. Question: Discuss the concept of inelastic demand in the context of necessities versus luxuries. Provide examples of each.

Solution: Inelastic demand implies that quantity demanded is less responsive to price changes. Necessities (e.g., basic food items) often have inelastic demand, while luxuries (e.g., high-end electronics) may have more elastic demand.

9. Question: How does the concept of time affect the elasticity of demand? Provide examples of goods with short-run and long-run elasticity differences.

Solution: In the short run, demand for certain goods may be inelastic (e.g., prescription medications). In the long run, consumers may find alternatives or adjust their habits, making the demand more elastic (e.g., electric cars as an alternative to traditional cars).

10. Question: Discuss the implications of technological advancements on the demand for certain products. How can technological innovations influence consumer preferences?

Solution: Technological advancements can lead to the introduction of new products or improvements to existing ones. This can significantly impact consumer preferences, influencing the demand for products that align with current technological trends.

11. Question: For a product, price was decreased from Rs 90 per unit to Rs 81 in order to attract more customers. It was perceived that demand for the product subsequently enlarged from 120 to 140 units. Calculate the price elasticity of demand

Solution: $E_P = \dfrac{\Delta Q}{\Delta P} \times \dfrac{P}{Q}$

$\Delta Q = 140 - 120 = 20$

$\Delta P = 81 - 90 = -9$

$E_P = \dfrac{20}{-9} \times \dfrac{90}{120} = -1.67$

The product has a relatively elastic demand.

12. Question: Find the income elasticity of demand for a consumer if his income rises from Rs 130 to Rs 250 and the quantity of a good bought by him increases from 35 units to 50 units.

Solution: $E_I = \dfrac{\Delta Q}{\Delta I} \times \dfrac{I}{Q}$

$\Delta Q = 50 - 35 = 15$

$\Delta I = 250 - 130 = 120$

$E_I = \dfrac{15}{120} \times \dfrac{130}{35} = 0.46$

13. Question: The quantity demanded of a product upsurges from 4000 units to 12,000 units due to increase in advertisement expenditure from Rs 3000 to Rs 15000. Find the promotional elasticity of demand.

Solution: $E_A = \frac{\Delta Q}{\Delta A} \times \frac{A}{Q}$

$\Delta Q = 12000 - 4000 = 8000$

$\Delta A = 15000 - 3000 = 12000$

$E_A = \frac{8000}{12000} \times \frac{3000}{4000} = 0.5$

C Code for Elasticity Calculation

```c
#include <stdio.h>

// Function to calculate elasticity
float calculateElasticity(float initialQuantity, float finalQuantity, float initialPrice, float finalPrice) {
    // Calculate percentage change in quantity and price
    float percentageChangeQuantity = ((finalQuantity - initialQuantity) / initialQuantity) * 100;
    float percentageChangePrice = ((finalPrice - initialPrice) / initialPrice) * 100;

    // Calculate elasticity using the formula
    float elasticity = percentageChangeQuantity / percentageChangePrice;

    return elasticity;
}

int main() {
    // Input values
    float initialQuantity, finalQuantity, initialPrice, finalPrice;

    // Get user input
    printf("Enter initial quantity: ");
    scanf("%f", &initialQuantity);

    printf("Enter final quantity: ");
    scanf("%f", &finalQuantity);

    printf("Enter initial price: ");
    scanf("%f", &initialPrice);

    printf("Enter final price: ");
    scanf("%f", &finalPrice);

    // Calculate and display elasticity
```

```
    float elasticity = calculateElasticity(initialQuantity, finalQuantity,
initialPrice, finalPrice);
    printf("Price Elasticity of Demand: %.2f\n", elasticity);

    return 0;
}
```

JAVA Code for Elasticity Calculation

```java
import java.util.Scanner;

public class ElasticityCalculator {

    // Function to calculate elasticity
    private static float calculateElasticity(float initialQuantity, float
finalQuantity, float initialPrice, float finalPrice) {
        // Calculate percentage change in quantity and price
        float    percentageChangeQuantity    =    ((finalQuantity    -
initialQuantity) / initialQuantity) * 100;
        float percentageChangePrice = ((finalPrice - initialPrice) /
initialPrice) * 100;

        // Calculate elasticity using the formula
        float    elasticity    =    percentageChangeQuantity    /
percentageChangePrice;

        return elasticity;
    }

    public static void main(String[] args) {
        // Input values
        float initialQuantity, finalQuantity, initialPrice, finalPrice;

        // Scanner for user input
        Scanner scanner = new Scanner(System.in);

        // Get user input
        System.out.print("Enter initial quantity: ");
        initialQuantity = scanner.nextFloat();

        System.out.print("Enter final quantity: ");
        finalQuantity = scanner.nextFloat();

        System.out.print("Enter initial price: ");
```

```java
initialPrice = scanner.nextFloat();

System.out.print("Enter final price: ");
finalPrice = scanner.nextFloat();

// Calculate and display elasticity
float elasticity = calculateElasticity(initialQuantity, finalQuantity,
initialPrice, finalPrice);
System.out.printf("Price    Elasticity    of    Demand:    %.2f\n",
elasticity);

// Close the scanner
scanner.close();
    }
}
```

PYTHON Code for Elasticity Calculation

```python
def calculate_elasticity(initial_quantity, final_quantity, initial_price, final_price):
    # Calculate percentage change in quantity and price
    percentage_change_quantity = ((final_quantity - initial_quantity) / initial_quantity) * 100
    percentage_change_price = ((final_price - initial_price) / initial_price) * 100

    # Calculate elasticity using the formula
    elasticity = percentage_change_quantity / percentage_change_price

    return elasticity


def main():
    # Input values
    initial_quantity = float(input("Enter initial quantity: "))
    final_quantity = float(input("Enter final quantity: "))
    initial_price = float(input("Enter initial price: "))
    final_price = float(input("Enter final price: "))

    # Calculate and display elasticity
    elasticity = calculate_elasticity(initial_quantity, final_quantity, initial_price, final_price)
    print(f"Price Elasticity of Demand: {elasticity:.2f}")


if __name__ == "__main__":
    main()
```

6 MONEY-VALUE OF MONEY

The concept of the money value of money refers to the idea that the purchasing power of money is not constant and can change over time. It is a fundamental aspect of monetary economics that explores how changes in the quantity of money, inflation, interest rates, and other economic factors impact the real value of money. In this extensive discussion, we will delve into the historical development of the money value of money, its key determinants, and its implications for individuals, businesses, and economies.

6.1 Historical Evolution

The understanding of the money value of money has evolved over centuries, reflecting the changing nature of economies and monetary systems. In ancient times, various commodities were used as money, such as gold, silver, and shells. The value of these commodities as money was influenced by their scarcity, durability, and other intrinsic qualities. The advent of metallic coins and later paper money marked significant milestones in the history of the money value of money. As economies transitioned from commodity money to representative and fiat money, the relationship between the quantity of money and its purchasing power became more complex.

6.2 Determinants of the Money Value of Money

Several factors contribute to the determination of the value of money. Understanding these factors is crucial for economists, policymakers, and individuals in assessing the real worth of money in different economic scenarios.

1. Inflation: Inflation is one of the primary determinants of the money value of money. When the general price level rises, the purchasing power of money decreases. Inflation erodes the real value of money, and individuals find that the same amount of money buys fewer goods and services over time.

2. Interest Rates: Interest rates play a crucial role in influencing the money value of money. Higher interest rates provide individuals and businesses with an incentive to hold onto money rather than spend it or invest it. Conversely, lower interest rates may encourage spending and investment, affecting the demand for money.

3. Economic Growth: The overall economic activity and growth rate of an economy can impact the time value of money. In growing economies, there is often increased demand for goods and services, potentially leading to higher prices and affecting the real value of money.

4. Government Policies: Monetary and fiscal policies implemented by governments and central banks can influence the money value of money. For example, a central bank's decision to increase or decrease the money supply can have significant effects on inflation and, consequently, the purchasing power of money.

5. Global Economic Factors: Factors such as exchange rates, international trade, and global economic conditions can also impact the time value of money. Changes in these factors can influence the prices of imported goods and services, contributing to inflation or deflation in a particular country.

6.3 Impact on Individuals and Businesses

The value of money has profound implications for individuals and businesses as they make financial decisions, plan for the future, and navigate economic uncertainties.

1. Purchasing Power:

Individuals experience the impact of the money value of money through changes in their purchasing power. Inflation reduces the purchasing power of money, affecting the standard of living and the ability to afford goods and services.

2. Investment Decisions:

Investors consider the money value of money when making investment decisions. Factors such as expected inflation and interest rates influence the real returns on investments. Understanding the dynamics of the money value of money is essential for making informed investment choices.

3. Cost of Borrowing:

Businesses and individuals assess the cost of borrowing in the context of the money value of money. Changes in interest rates and inflation can affect the real cost of loans, impacting borrowing decisions and overall financial planning.

4. Income and Wages:

Changes in the money value of money can affect real wages and income. If wages do not keep pace with inflation, individuals may find their real income decreasing, impacting their overall financial well-being.

6.4 Monetary Policies and Central Banking

Central banks play a crucial role in managing the value of money through monetary policies. The primary tools at the disposal of central banks include open market operations, reserve requirements, and discount rates.

1. Open Market Operations:

Central banks conduct open market operations to buy or sell government securities. These transactions influence the money supply, affecting interest rates and, consequently, the money value of money.

2. Reserve Requirements:

Central banks set reserve requirements, determining the amount of money that banks must hold in reserve. Adjustments to reserve requirements can impact the money supply and lending capacity of commercial banks.

3. Discount Rates:

Central banks use the discount rate to influence the cost of borrowing for commercial banks. Changes in the discount rate can impact interest rates throughout the economy, affecting spending and investment decisions.

4. Forward Guidance:

Central banks often provide forward guidance on their future monetary policy intentions. This guidance can influence expectations about future interest rates and inflation, shaping economic behavior.

6.5 Challenges and Criticisms

While the concept of the money value of money is fundamental to economic analysis, it is not without challenges and criticisms. One notable criticism is that traditional economic models often assume a direct and linear relationship between the quantity of money and its value, overlooking complexities in the real world.

1. Lag Effects:

The effects of monetary policy on the money value of money may not be immediate. There can be lags between policy actions and their impact on inflation and interest rates.

2. Uncertainty and Expectations:

Human expectations and perceptions play a significant role in shaping economic behavior. Unpredictable events and shifts in expectations can lead to outcomes that deviate from standard economic predictions.

3. Global Interconnectedness:

In an interconnected global economy, domestic monetary policies can be influenced by international factors. Exchange rate movements and global economic conditions add layers of complexity to the determination of the money value of money.

4. Non-Traditional Monetary Policies:

The use of unconventional monetary policies, such as quantitative easing, has raised questions about their long-term impact on the money value of money and the potential unintended consequences.

The time value of money is a dynamic and multifaceted concept that lies at the heart of monetary economics. It encapsulates the intricate relationship between the quantity of money, inflation, interest rates, and various economic factors. As economies evolve and financial systems become increasingly complex, the understanding of the time value of money continues to be refined and adapted.

For individuals, businesses, and policymakers, recognizing the nuances of the money value of money is essential for making informed decisions, whether it be in managing personal finances, making investment choices, or formulating monetary policies. The ongoing interplay between economic variables and the money value of money underscores the need for continuous research, analysis, and adaptability in the face of an ever-changing economic landscape.

6.6 Calculation Technique

The time value of money (TVM) is a fundamental concept in finance that recognizes the idea that a sum of money today has a different value than the same sum in the future. This concept is based on the premise that money has the potential to earn returns or incur costs over time due to factors such as interest rates, inflation, and opportunity costs. The TVM is crucial for making financial decisions, comparing investment opportunities, and evaluating the cost of capital.

The mathematical formula for calculating the time value of money is often represented by the following basic equations:

1. Future Value (FV) Formula:

The future value represents the value of a sum of money at a future date, taking into account compound interest. The formula is given by:

$$FV = PV \times (1 + r)^n$$

Where:
- FV is the future value of the investment or loan.
- PV is the present value or initial principal amount.
- r is the interest rate per compounding period (expressed as a decimal).
- n is the number of compounding periods.

2. Present Value (PV) Formula:

The present value is the current worth of a sum of money to be received or paid in the future, discounted at a certain interest rate. The formula is given by:

$$PV = \frac{FV}{(1 + r)^n}$$

Where:
- PV is the present value of the investment or loan.
- FV is the future value of the money.
- r is the interest rate per compounding period (expressed as a decimal).
- n is the number of compounding periods.

3. Net Present Value (NPV) Formula:

Net Present Value is used to evaluate the profitability of an investment by comparing the present value of expected cash inflows with the present value of expected cash outflows. The formula is given by:

$$NPV = \sum_{t=0}^{T} \frac{CF_t}{(1+r)^t}$$

Where:
- NPV is the net present value.
- T is the total number of time periods.
- CF_t is the net cash inflow or outflow during time period t.
- r is the discount rate.

4. Discounted Cash Flow (DCF) Formula:

DCF is used to estimate the value of an investment based on its expected future cash flows. The formula is given by:

$$DCF = \frac{CF_1}{(1+r)^1} + \frac{CF_2}{(1+r)^2} + \frac{CF_3}{(1+r)^3} + \cdots + \frac{CF_n}{(1+r)^n}$$

where:
- DCF is the discounted cash flow.
- CF_t is the net cash flow during time period t.
- r is the discount rate.
- n is the number of time periods.

These formulas provide a mathematical framework for understanding and calculating the time value of money in various financial contexts. The TVM concept is critical for financial decision-making, investment analysis, and assessing the opportunity cost of allocating resources over time.

Suggested Questions with Solutions

1. Question: Define and explain the concept of the time value of money. How does it impact financial decision-making?

Solution: The time value of money (TVM) is the idea that the value of a sum of money changes over time due to factors like interest, inflation, and opportunity costs. Financial decision-making considers TVM through concepts such as present value, future value, and discounted cash flows. It is fundamental in evaluating investments, loans, and the cost of capital.

2. Question: Describe the factors that influence the money value of money. How does inflation, interest rates, and economic conditions impact the purchasing power of currency?

Solution: The money value of money is influenced by several factors. Inflation erodes purchasing power, causing money to lose value over time. Changes in interest rates affect the opportunity cost of holding money. Economic conditions, such as recessions or growth, can impact overall purchasing power and the value of money.

3. Question: Explain the relationship between inflation and the money value of money. How does the purchasing power of money change during inflationary periods?

Solution: Inflation reduces the purchasing power of money. As prices rise, the same amount of money buys fewer goods and services. This is because inflation erodes the real value of currency. Understanding this relationship is crucial for individuals and businesses to adapt their financial strategies during inflationary periods.

4. Question: Discuss the role of central banks in managing the money value of money. How do monetary policies influence

inflation, interest rates, and currency stability?

Solution: Central banks play a crucial role in managing the money value of money through monetary policies. By adjusting interest rates, conducting open market operations, and controlling the money supply, central banks aim to maintain price stability, influence inflation rates, and ensure the stability of the currency.

5. Question: Explore the concept of opportunity cost in the context of the money value of money. How does the decision to hold money instead of investing impact overall wealth accumulation?

Solution: Opportunity cost refers to the potential benefits foregone when one choice is made over another. Holding money instead of investing incurs an opportunity cost, as the money could have earned returns. Understanding and quantifying this opportunity cost is crucial for making informed financial decisions that maximize wealth accumulation over time.

6. Question: You would like to buy a house that is currently on the market at $80,000, but you cannot afford it right now. However, you think that you would be able to buy it after 5 years. If the expected inflation rate as applied to the price of this house is 5% per year, what is its expected price after four years?

Solution: Future Value (FV): $PV(1+r)^n$
$$= 80000(1+0.05)^5$$
$$= \$1,02,102.525$$

7. Question: Jamini has deposited $4,000 in a money market account with a variable interest rate. The account compounds the interest monthly. Jamini expects the interest rate to remain at 9% annually for the first 4 months, at 10% annually for the next 4 months, and then back to 11% annually for the next 4 months. Find the total amount in this account after 12 months.

Solution: FV = $4000(1+0.09/12)^4(1+0.1/12)^4(1+0.11/12)^4$
= $4414.17

8. Question: You decide to put $10,000 in a money market fund that pays interest at the annual rate of 8.4%, compounding it monthly. You plan to take the money out after one year and pay the income tax on the interest earned. You are in the 12% tax bracket. Find the total amount available to you after taxes.

Solution: The monthly interest rate is $.084/12 = .007$. Using it as the growth rate, the future value of money after twelve months is:
$$FV = 10000(1+0.007)^{12} = 10,873.11$$
The interest earned = $10,873.11 - 10,000 = $873.11. You have to pay 12% tax on this amount. Thus, after paying taxes, it becomes
= $873.11(1 - .12) = $768.33
Total amount available after 12 months = $10,000 + 768.33 = $10,768.33.

C Code for Calculating Time Value of Money

```c
#include <stdio.h>
#include <math.h>

// Function to calculate Present Value (PV)
double presentValue(double futureValue, double interestRate, int periods) {
    return futureValue / pow(1 + interestRate, periods);
}

// Function to calculate Future Value (FV)
double futureValue(double presentValue, double interestRate, int periods) {
    return presentValue * pow(1 + interestRate, periods);
}

// Function to calculate Net Present Value (NPV)
double netPresentValue(double cashFlows[], double discountRate, int numPeriods) {
    double npv = 0.0;
    for (int i = 0; i < numPeriods; ++i) {
        npv += cashFlows[i] / pow(1 + discountRate, i + 1);
    }
    return npv;
}

int main() {
    // Example usage
    double futureValueAmount = 1000.0;
    double interestRate = 0.05;
    int periods = 3;

    // Calculate Present Value
    double presentVal = presentValue(futureValueAmount, interestRate, periods);
    printf("Present Value: %.2f\n", presentVal);
```

```
    // Calculate Future Value
    double futureVal = futureValue(presentVal, interestRate,
periods);
    printf("Future Value: %.2f\n", futureVal);

    // Example for Net Present Value
    double cashFlows[] = {-1000, 300, 300, 300, 300};
    int numPeriodsNPV = sizeof(cashFlows) /
sizeof(cashFlows[0]);

    double discountRateNPV = 0.1;
    double npv = netPresentValue(cashFlows, discountRateNPV,
numPeriodsNPV);
    printf("Net Present Value: %.2f\n", npv);

    return 0;
}
```

JAVA Code for Calculating Time Value of Money

```java
import java.util.Scanner;

public class TimeValueOfMoneyCalculator {

    // Function to calculate Present Value (PV)
    private static double presentValue(double futureValue, double
interestRate, int periods) {
        return futureValue / Math.pow(1 + interestRate, periods);
    }

    // Function to calculate Future Value (FV)
    private static double futureValue(double presentValue, double
interestRate, int periods) {
        return presentValue * Math.pow(1 + interestRate, periods);
    }

    // Function to calculate Net Present Value (NPV)
    private static double netPresentValue(double[] cashFlows,
double discountRate) {
        double npv = 0.0;
        for (int i = 0; i < cashFlows.length; i++) {
            npv += cashFlows[i] / Math.pow(1 + discountRate, i + 1);
        }
        return npv;
    }

    public static void main(String[] args) {
        // Example usage
        Scanner scanner = new Scanner(System.in);

        System.out.print("Enter future value: ");
        double futureValueAmount = scanner.nextDouble();

        System.out.print("Enter interest rate (in decimal): ");
        double interestRate = scanner.nextDouble();
```

```java
        System.out.print("Enter number of periods: ");
        int periods = scanner.nextInt();

        // Calculate Present Value
        double presentVal = presentValue(futureValueAmount,
interestRate, periods);
        System.out.printf("Present Value: %.2f\n", presentVal);

        // Calculate Future Value
        double futureVal = futureValue(presentVal, interestRate,
periods);
        System.out.printf("Future Value: %.2f\n", futureVal);

        // Example for Net Present Value
        System.out.print("Enter number of cash flow periods: ");
        int numPeriodsNPV = scanner.nextInt();

        double[] cashFlows = new double[numPeriodsNPV];
        for (int i = 0; i < numPeriodsNPV; i++) {
            System.out.printf("Enter cash flow for period %d: ", i + 1);
            cashFlows[i] = scanner.nextDouble();
        }

        System.out.print("Enter discount rate (in decimal): ");
        double discountRateNPV = scanner.nextDouble();

        double npv = netPresentValue(cashFlows,
discountRateNPV);
        System.out.printf("Net Present Value: %.2f\n", npv);

        // Close the scanner
        scanner.close();
    }
}
```

PYTHON Code for Calculating Time Value of Money

```python
def present_value(future_value, interest_rate, periods):
    return future_value / (1 + interest_rate) ** periods

def future_value(present_value, interest_rate, periods):
    return present_value * (1 + interest_rate) ** periods

def net_present_value(cash_flows, discount_rate):
    npv = sum(cash_flow / (1 + discount_rate) ** (i + 1) for i,
cash_flow in enumerate(cash_flows))
    return npv

# Example usage
future_value_amount = float(input("Enter future value: "))
interest_rate = float(input("Enter interest rate (in decimal): "))
periods = int(input("Enter number of periods: "))

# Calculate Present Value
present_val = present_value(future_value_amount, interest_rate,
periods)
print(f"Present Value: {present_val:.2f}")

# Calculate Future Value
future_val = future_value(present_val, interest_rate, periods)
print(f"Future Value: {future_val:.2f}")

# Example for Net Present Value
num_periods_npv = int(input("Enter number of cash flow periods:
"))
cash_flows = [float(input(f"Enter cash flow for period {i + 1}: "))
for i in range(num_periods_npv)]
discount_rate_npv = float(input("Enter discount rate (in decimal):
"))

npv = net_present_value(cash_flows, discount_rate_npv)
print(f"Net Present Value: {npv:.2f}")
```

7 THE QUANTITY THEORY OF MONEY: UNDERSTANDING THE DYNAMICS OF INFLATION AND DEFLATION

7.1 Introduction

The Quantity Theory of Money (QTM) is a fundamental concept in monetary economics that seeks to explain the relationship between the quantity of money in circulation and the level of prices in an economy. Developed over centuries and refined by economists such as Irving Fisher and Milton Friedman, the Quantity Theory serves as a cornerstone in understanding the dynamics of inflation and deflation. In this comprehensive exploration, we will delve into the components of the Quantity Theory, the implications for price movements, and the contrasting phenomena of inflation and deflation.

7.2 Components of the Quantity Theory of Money

The Quantity Theory of Money is often expressed in the form of an equation known as the Quantity Equation, which provides a framework for understanding the relationship between the money supply, the velocity of money, and the price level. The equation is expressed as:

$$M \times V = P \times T$$

where:
- M represents the money supply.
- V represents the velocity of money (the average number of times a unit of money is spent in a specific time period).
- P represents the price level.
- T represents the volume of transactions.

According to the Quantity Theory, changes in the money supply M will directly impact the price level P, assuming velocity V and the volume of transactions T remain relatively constant. This theory implies that an increase in the money supply, if not matched by an increase in the volume of transactions, will lead to a proportional increase in prices.

7.3 Inflation

Inflation is the sustained increase in the general price level of goods and services in an economy over time. It is a key economic phenomenon that affects consumers, businesses, and policymakers. The Quantity Theory of Money provides insights into the factors that contribute to inflation.

7.3.1 Causes of Inflation
- **Demand-Pull Inflation:** Occurs when aggregate demand exceeds aggregate supply, leading to increased prices. This often happens in growing economies with high consumer confidence and spending.

- **Cost-Push Inflation:** Arises from increased production costs, such as higher wages or increased prices of raw materials. These cost increases are passed on to consumers in the form of higher prices.

- **Built-In Inflation:** Also known as wage-price inflation, it occurs when workers demand higher wages, and businesses pass these higher costs on to consumers in the form of higher prices.

- **Monetary Inflation:** Linked to an increase in the money supply. According to the Quantity Theory of Money, if the money supply grows faster than the real output of goods and services, it can contribute to inflation.

7.4 Deflation

Deflation is the opposite of inflation—it is a sustained decrease in the general price level of goods and services. While less common than inflation, deflation can have significant economic implications and challenges.

7.4.1 Causes of Deflation

- **Demand-Deficient Deflation:** Occurs when aggregate demand is insufficient to absorb the existing supply of goods and services. This can lead to reduced prices as businesses lower prices to stimulate demand.

- **Productivity-Enhancing Deflation:** Can occur in economies experiencing rapid technological advancements and increased productivity. While falling prices may benefit consumers, it poses challenges for businesses in maintaining profitability.

- **Debt-Deflation:** Arises when the burden of existing debt becomes heavier due to falling prices. As the real value of money increases, borrowers may struggle to repay debt, leading to economic contraction.

- **Monetary Deflation:** Linked to a decrease in the money supply. If the money supply contracts, there is less money available to support economic transactions, potentially leading to deflation.

7.5 The Phillips Curve: Trade-Off Between Inflation and Unemployment

While the Quantity Theory of Money provides insights into the relationship between the money supply and inflation, the Phillips Curve introduces another dimension by examining the trade-off between inflation and unemployment. Developed by economist A.W. Phillips, the Phillips Curve suggests an inverse relationship between inflation and unemployment in the short run. According to the Phillips Curve, policymakers face a trade-off between achieving low inflation and low unemployment.

However, the Phillips Curve has been subject to criticism, particularly as economists recognized that the trade-off is not always stable and that there can be periods of both high inflation and high unemployment, as witnessed during episodes of stagflation.

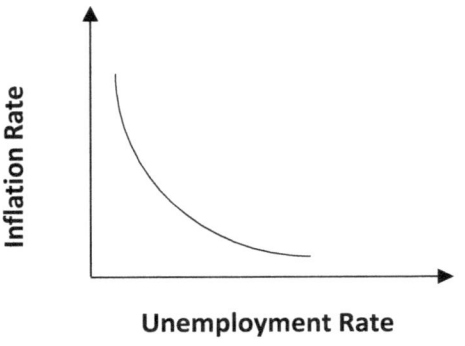

Fig. 3: The Phillips Curve

7.6 Policy Implications

Understanding the dynamics of inflation and deflation has significant policy implications for central banks and governments. Monetary and fiscal policies are key tools used to manage and mitigate the impact of these economic phenomena.

- **Monetary Policy:** Central banks use tools such as interest rates and open market operations to control the money supply and influence inflation. By adjusting interest rates, central banks can impact borrowing and spending, influencing the level of economic activity and inflation.

- **Fiscal Policy:** Governments can use fiscal measures, such as taxation and government spending, to manage demand in the economy. During periods of high inflation, policymakers may adopt contractionary fiscal policies to cool off an overheated economy. In contrast, during deflationary periods, expansionary fiscal policies may be implemented to stimulate economic activity.

7.7 Challenges and Criticisms

While the Quantity Theory of Money and related economic models provide valuable insights, they are not without challenges and criticisms. Some of the key criticisms include:

- **Assumption of Constant Velocity:** The Quantity Theory assumes that the velocity of money remains relatively constant. However, velocity can fluctuate due to changes in economic conditions and financial innovations.

- **Real-World Complexity:** The real world is characterized by complexity and a multitude of factors influencing inflation and deflation. Economic phenomena are often influenced by a combination of factors, and simplistic models may not capture the full dynamics.

- **Expectations and Psychology:** Expectations and psychological factors play a significant role in shaping economic behavior. Consumer and business expectations can influence spending, saving, and investment decisions, impacting inflation and deflation.

7.8 Conclusion

The Quantity Theory of Money provides a foundational framework for understanding the relationship between the money supply and the price level, offering insights into the dynamics of inflation and deflation. Inflation, characterized by a sustained increase in prices, has various causes ranging from demand-pull to cost-push factors. Deflation, marked by a sustained decrease in prices, poses challenges such as debt-deflation and demand-deficient deflation.

Understanding these economic phenomena is crucial for policymakers, businesses, and individuals in making informed decisions. Central banks and governments implement monetary and fiscal policies to manage inflation and deflation, recognizing the delicate balance required to achieve economic stability.

While economic models and theories provide valuable guidance, it's essential to acknowledge the complexity of real-world economic systems. The Quantity Theory of Money, the Phillips Curve, and related frameworks contribute to a holistic understanding of inflation and deflation, but continuous research and adaptability are necessary to navigate the ever-changing landscape of the global economy.

Suggested Questions with Solutions

1. Question: Explain the Quantity Theory of Money and its key components. How does the theory depict the relationship between money supply, velocity, and the price level?

Solution: The Quantity Theory of Money is represented by the equation $M \times V = P \times T$, where M is the money supply, V is the velocity of money, P is the price level, and T is the volume of transactions. The theory posits that changes in the money supply will proportionally affect the price level, assuming velocity and transaction volume remain relatively constant.

2. Question: Discuss the causes and consequences of inflation. How do demand-pull, cost-push, and built-in inflation contribute to the overall rise in prices?

Solution: Inflation can result from various factors:
  - **Demand-Pull Inflation:** Caused by increased aggregate demand exceeding aggregate supply.
  - **Cost-Push Inflation:** Arises from increased production costs, such as higher wages or raw material prices.
  - **Built-In Inflation:** Occurs when workers demand higher wages, leading to increased production costs.

  Consequences of inflation include reduced purchasing power, uncertainty, and challenges for fixed-income earners.

3. Question: Examine the concept of deflation and its causes. How does deflation differ from inflation, and what are the potential consequences of a sustained decrease in the price level?

Solution: Deflation is the sustained decrease in the general price level. Causes include reduced aggregate demand, productivity gains, or debt-deflation. Unlike inflation, deflation can lead to increased real debt burdens, postponed spending, and economic contraction, posing challenges for policymakers.

4. Question: Explore the Phillips Curve and its implications for the trade-off between inflation and unemployment. How does the short-run relationship depicted by the Phillips Curve influence economic policy decisions?

Solution: The Phillips Curve suggests an inverse relationship between inflation and unemployment in the short run. Policymakers often face a trade-off where reducing inflation may lead to increased unemployment and vice versa. However, the stability of this trade-off is subject to change, as demonstrated during episodes of stagflation.

5. Question: Explain how monetary and fiscal policies can be used to address inflation and deflation. Provide examples of policy measures that central banks and governments may employ during economic challenges.

Solution:
  - **Monetary Policy:** Central banks can adjust interest rates, conduct open market operations, or set reserve requirements to influence money supply and inflation.
  - **Fiscal Policy:** Governments can use taxation and spending to manage demand in the economy. During inflation, contractionary fiscal policies may be employed, while expansionary policies could be used during deflationary periods.

6. Question: How much life insurance should a person buy if he wants to leave enough money to his family, so they receive $30,000 per year in interest, of consent Year 0 value dollars? The interest rate expected from banks is 10%, while the inflation rate is expected to be 5% per year.

Solution: The actual (effective) rate that the family will be getting is

$$i' = \frac{i-f}{1+f} = \frac{0.1-0.05}{1+0.05} = 0.0476 = 4.76\%$$

To calculate P, n = ∞ (capitalized cost)

$$P = \frac{A}{i'} = \frac{\$30,000}{0.0476} = \$6,30,000$$

Therefore, he needs to buy about $6,30,000 of life insurance.

C Code to Calculate Inflated Price for 'n' Years

```c
#include <stdio.h>
#include <math.h>

// Function to calculate inflated price for 'n' years
float calculateInflatedPrice(float originalPrice, float inflationRate, int years) {
    // Calculate the inflated price using the formula: Inflated Price
= Original Price * (1 + Inflation Rate)^Years
    float inflatedPrice = originalPrice * pow(1 + inflationRate, years);
    return inflatedPrice;
}

int main() {
    // Variables to store original price, inflation rate, and number of years
    float originalPrice, inflationRate;
    int years;

    // Get user input for original price
    printf("Enter the original price: $");
    scanf("%f", &originalPrice);

    // Get user input for inflation rate
    printf("Enter the annual inflation rate (as a decimal): ");
    scanf("%f", &inflationRate);

    // Get user input for the number of years
    printf("Enter the number of years: ");
    scanf("%d", &years);

    // Check if the inflation rate is negative (indicating deflation)
    if (inflationRate < 0) {
        printf("Warning: Negative inflation rate. This indicates deflation.\n");
```

```
    }

    // Calculate and display the inflated price for 'n' years
    float    inflatedPrice    =    calculateInflatedPrice(originalPrice,
inflationRate, years);
    printf("The inflated price after %d years is: $%.2f\n", years,
inflatedPrice);

    return 0;
}
```

JAVA Code to Calculate Inflated Price for 'n' Years

```java
import java.util.Scanner;

public class InflatedPriceCalculator {

    // Function to calculate inflated price for 'n' years
    public static double calculateInflatedPrice(double originalPrice,
    double inflationRate, int years) {
        // Calculate the inflated price using the formula: Inflated Price
    = Original Price * (1 + Inflation Rate)^Years
        double inflatedPrice = originalPrice * Math.pow(1 +
    inflationRate, years);
        return inflatedPrice;
    }

    public static void main(String[] args) {
        // Variables to store original price, inflation rate, and number
    of years
        double originalPrice, inflationRate;
        int years;

        // Scanner for user input
        Scanner scanner = new Scanner(System.in);

        // Get user input for original price
        System.out.print("Enter the original price: $");
        originalPrice = scanner.nextDouble();

        // Get user input for inflation rate
        System.out.print("Enter the annual inflation rate (as a
    decimal): ");
        inflationRate = scanner.nextDouble();

        // Get user input for the number of years
        System.out.print("Enter the number of years: ");
        years = scanner.nextInt();
```

```
// Check if the inflation rate is negative (indicating deflation)
if (inflationRate < 0) {
    System.out.println("Warning: Negative inflation rate. This
indicates deflation.");
}

// Calculate and display the inflated price for 'n' years
double inflatedPrice = calculateInflatedPrice(originalPrice,
inflationRate, years);
    System.out.printf("The inflated price after %d years is:
$%.2f%n", years, inflatedPrice);

// Close the scanner
    scanner.close();
    }
}
```

PYTHON Code to Calculate Inflated Price for 'n' Years

```python
def calculate_inflated_price(original_price, inflation_rate, years):
    # Calculate the inflated price using the formula: Inflated Price = Original Price * (1 + Inflation Rate)^Years
    inflated_price = original_price * (1 + inflation_rate) ** years
    return inflated_price

# Get user input for original price
original_price = float(input("Enter the original price: $"))

# Get user input for inflation rate
inflation_rate = float(input("Enter the annual inflation rate (as a decimal): "))

# Get user input for the number of years
years = int(input("Enter the number of years: "))

# Check if the inflation rate is negative (indicating deflation)
if inflation_rate < 0:
    print("Warning: Negative inflation rate. This indicates deflation.")

# Calculate and display the inflated price for 'n' years
inflated_price = calculate_inflated_price(original_price, inflation_rate, years)
print(f"The inflated price after {years} years is: ${inflated_price:.2f}")
```

8 BANKING: ROLE OF COMMERCIAL BANKS, CREDIT, AND ITS IMPORTANCE IN INDUSTRIAL FUNCTIONING – SOURCE OF FINANCE

8.1 Introduction

𝕭anking is a cornerstone of modern economic systems, playing a crucial role in facilitating financial transactions, supporting economic activities, and contributing to the overall development of a nation. Commercial banks, as key entities in the banking sector, hold a pivotal position in the financial landscape. This comprehensive exploration delves into the multifaceted roles of commercial banks, the significance of credit, and how these elements serve as vital sources of finance for industrial development.

8.2 The Role of Commercial Banks

Commercial banks serve as key financial intermediaries, playing a crucial role in the economic functioning of a country. These institutions are at the forefront of providing a range of financial services to individuals, businesses, and the government. One primary function is the mobilization of funds through various deposit products, including savings accounts, fixed deposits, and current accounts. By attracting deposits, commercial banks accumulate the capital needed to extend loans to borrowers. This credit intermediation role is fundamental to economic growth, as it enables individuals and businesses to access the capital necessary for investments, expansions, and daily operations.

Commercial banks also facilitate payments and transactions within the economy by offering services such as check clearing, electronic funds transfers, and the issuance of credit and debit cards. They provide a secure platform for individuals and businesses to store valuables and important documents through safe deposit boxes. Additionally, commercial banks contribute to the global interconnectedness of financial markets by engaging in foreign exchange activities, supporting international trade, and offering trade finance services.

Beyond traditional banking services, many commercial banks provide advisory and wealth management services, assisting clients with financial planning, investments, and retirement strategies. As financial intermediaries, commercial banks help manage liquidity in the financial system, ensuring a balance between the supply and demand for funds. Overall, the multifaceted roles of commercial banks make them integral to the economic landscape, fostering economic growth, stability, and financial inclusion.

8.2.1. Financial Intermediation
Commercial banks function as intermediaries between depositors and borrowers. They accept deposits from individuals and businesses and use these funds to extend loans. This financial intermediation is fundamental for the efficient allocation of resources in an economy.

8.2.2 Depository Services
One of the primary functions of commercial banks is to provide a safe and secure place for individuals and businesses to deposit their money. This function enhances financial stability by reducing the risks associated with holding large sums of cash.

8.2.3 Lending and Credit Creation
Commercial banks are major lenders, providing credit to various

sectors of the economy. Through the process of credit creation, banks leverage the initial deposits to extend loans, fostering economic growth and development.

8.2.4 Payment Services

Commercial banks facilitate transactions by providing payment services, including issuing checks, electronic fund transfers, and facilitating international trade through letters of credit. These services contribute to the efficiency of the payment system, promoting economic activities.

8.2.5 Investment Banking Services

In addition to traditional banking activities, many commercial banks engage in investment banking services. This includes underwriting securities, facilitating mergers and acquisitions, and providing advisory services. Investment banking plays a vital role in capital markets and corporate finance.

8.3 Credit and Its Importance in Industrial Functioning

Credit plays a pivotal role in the functioning of industrial enterprises by providing them with the necessary financial resources for growth, investment, and day-to-day operations. Industrial activities often require substantial capital, and credit serves as a means for businesses to access funds beyond their immediate financial capacity. Companies utilize credit for a variety of purposes, including purchasing equipment, expanding facilities, financing research and development, and managing working capital needs. The availability of credit enables businesses to make strategic investments that contribute to increased production, innovation, and overall competitiveness. Moreover, credit facilitates the smooth flow of commerce by allowing companies to extend payment terms to suppliers and customers, supporting trade relationships. However, the responsible use of credit is crucial, and businesses must carefully manage their debt to ensure sustainability and avoid financial strain. Overall, credit is an essential tool that empowers industrial enterprises to pursue opportunities, navigate challenges, and contribute to economic development.

8.3.1 Definition of Credit

Credit is a financial arrangement where a borrower receives funds from a lender with the obligation to repay the principal amount along with interest over a specified period. Credit is a vital component of economic activity, enabling individuals and businesses to invest, expand, and meet short-term financial needs.

8.3.2 Importance of Credit in Industrial Functioning

a. Capital Formation:

Credit is a catalyst for capital formation, especially in the industrial sector. Industries often require substantial funds for machinery, technology, and infrastructure. Credit enables businesses to acquire the necessary capital for expansion and modernization.

b. Business Expansion:

Credit facilitates business expansion by providing the financial resources needed for increasing production capacity, entering new markets, and implementing innovation. Small and medium-sized enterprises (SMEs) particularly benefit from credit to fuel their growth.

c. Research and Development:

Industries involved in innovation and research heavily depend on credit for funding their projects. Credit supports the development of new technologies, processes, and products, contributing to industrial progress.

d. Working Capital Management:

Credit is essential for managing day-to-day operations and addressing short-term financial requirements. Industries often utilize credit to maintain sufficient working capital, ensuring smooth production and business continuity.

e. Job Creation:

Access to credit enables businesses to expand their operations, leading to increased job opportunities. As industries grow, they require additional workforce, contributing to employment generation and economic development.

8.4 Source of Finance

Sources of finance refer to the various means through which businesses and individuals secure funds for their activities and financial needs. These sources can be broadly categorized into two types: internal and external. Internal sources include funds generated from within the organization, such as retained earnings, where profits are reinvested back into the business. External sources encompass a wide range of options, such as debt financing through loans or bonds, equity financing by issuing shares, and alternative sources like crowdfunding or venture capital. The choice of financing source depends on factors such as the nature of the project, the cost of capital, risk tolerance, and the financial structure of the entity. Diversifying sources of finance is often a prudent strategy for reducing risk and ensuring financial stability. The selection of an appropriate mix of internal and external sources is a critical aspect of financial management, contributing to the sustainability and growth of businesses.

8.4.1 Debt Financing

Commercial banks are significant sources of debt financing. Businesses obtain loans from banks for various purposes, including capital investment, working capital needs, and expansion plans. Debt financing allows companies to access funds without diluting ownership.

8.4.2 Trade Credit

Trade credit is an essential source of finance, especially in the industrial sector. It involves the extension of credit terms by suppliers to buyers. This form of credit allows businesses to acquire goods and services on credit, providing a valuable source of short-term financing.

8.4.3 Equity Financing

While commercial banks primarily provide debt financing, equity financing is another crucial source of funds for industries. This involves issuing shares to investors, raising capital in exchange for ownership. Equity financing is often used for substantial capital-intensive projects.

8.4.4 Government Funding and Subsidies

Governments play a role in supporting industrial development by providing funding and subsidies. These funds may be channeled through commercial banks, offering favorable terms to industries engaged in priority sectors, research and development, or sustainable practices.

8.4.5 Private Equity and Venture Capital

Industries, particularly startups and high-growth companies, may seek funding from private equity firms and venture capitalists. These investors provide capital in exchange for ownership stakes, supporting innovative ventures and industries with high growth potential.

8.5 Challenges and Risks in Banking and Credit

The banking and credit industry faces a multitude of challenges and risks that require constant vigilance and strategic management. One significant challenge is the dynamic nature of the global financial landscape, characterized by economic uncertainties, geopolitical tensions, and regulatory changes. These factors can impact interest rates, creditworthiness, and the overall stability of financial institutions. Credit risk is a pervasive concern, as banks must assess and manage the likelihood of borrowers defaulting on loans. Market risk, including fluctuations in interest rates and foreign exchange rates, poses another challenge, affecting the value of financial instruments in banks' portfolios. Technological advancements, while providing opportunities, also introduce cybersecurity risks, as the sector becomes increasingly reliant on digital platforms. Compliance with evolving regulatory frameworks is an ongoing challenge, requiring banks to adapt to new standards while ensuring operational efficiency. Additionally, macroeconomic factors such as inflation, unemployment, and economic downturns can impact borrowers' ability to repay loans, posing systemic risks to the banking industry. Effective risk management practices, robust internal controls, and adaptability to the evolving financial landscape are crucial for banks to navigate these challenges and sustain their financial health.

8.5.1 Credit Risk

One of the primary challenges in banking is credit risk, the risk that borrowers may fail to repay their loans. Commercial banks employ rigorous risk assessment and credit scoring mechanisms to minimize this risk but must constantly navigate the balance between extending credit and managing potential defaults.

8.5.2 Interest Rate Risk

Commercial banks are exposed to interest rate risk, given that the interest rates they pay on deposits may differ from the rates they earn on loans. Changes in interest rates, influenced by economic conditions and monetary policy, can impact a bank's profitability.

8.5.3 Economic Downturns

During economic downturns, industries may face challenges in repaying loans, leading to an increase in non-performing assets for banks. The cyclicality of economies poses risks to both the banking sector and industries dependent on credit for expansion.

8.5.4 Regulatory Compliance

The banking sector operates under strict regulatory frameworks to ensure stability and protect depositors. Adhering to these regulations can be challenging for banks, and failure to comply may result in legal and financial consequences.

8.5.5 Technological Disruptions

The rise of financial technology (fintech) poses both opportunities and challenges for the banking sector. While technological advancements enhance efficiency, they also introduce new risks related to cybersecurity, data privacy, and competition from non-traditional financial service providers.

8.6 Future Trends and Innovations

Future trends and innovations are poised to reshape various aspects of society, technology, and business. In technology, the advent of artificial intelligence (AI) and machine learning is expected to drive unprecedented advancements in automation, data analytics, and decision-making processes. The Internet of Things (IoT) will further connect devices and systems, creating a more interconnected and efficient world. Sustainable technologies, renewable energy sources, and eco-friendly practices are gaining prominence as the global community intensifies efforts to address environmental challenges. In healthcare, precision medicine and gene editing hold the promise of personalized treatments and breakthroughs in disease prevention. Blockchain technology is transforming industries by enhancing security and transparency in transactions. Augmented reality (AR) and virtual reality (VR) are shaping new ways of communication, education, and entertainment. Future trends also encompass shifts in work patterns, with the rise of remote work and the gig economy. These innovations reflect an era of rapid change and present both opportunities and challenges as societies adapt to the evolving landscape.

8.6.1 Digital Banking

The future of banking is increasingly digital. Digital banking services, including online banking, mobile apps, and digital payment systems, are transforming the way customers interact with banks. This trend enhances accessibility and efficiency in financial transactions.

8.6.2 Blockchain and Cryptocurrencies

Emerging technologies like blockchain have the potential to revolutionize banking processes, particularly in areas such as secure and transparent transaction processing. Cryptocurrencies may offer alternative forms of finance, challenging traditional banking models.

8.6.3 Sustainable Finance

There is a growing emphasis on sustainable finance, with banks considering environmental, social, and governance (ESG) factors in their lending practices. This trend reflects a shift towards responsible banking and financing projects that align with sustainable development goals.

8.6.4 Open Banking

Open banking initiatives promote collaboration between traditional banks and third-party financial service providers. This approach allows customers to access a broader range of financial products and services, fostering competition and innovation.

8.6.5 Artificial Intelligence (AI) and Big Data

Banks are increasingly leveraging AI and big data analytics to enhance decision-making processes, risk management, and customer experiences. These technologies enable banks to analyze vast amounts of data to identify trends, manage risks, and personalize services.

8.7 Conclusion

In conclusion, commercial banks are linchpins in the financial infrastructure, serving as intermediaries, depositories, and providers of essential financial services. Credit, as facilitated by these banks, plays a pivotal role in industrial functioning, supporting capital formation, business expansion, and overall economic development. The sources of finance for industries are diverse, ranging from traditional debt financing provided by commercial banks to equity financing, trade credit, and government support.

While the banking sector and credit are integral to economic progress, they face challenges and risks, including credit risk, interest rate fluctuations, and regulatory compliance. The evolving landscape of banking includes future trends and innovations such as digital banking, blockchain, sustainable finance, open banking, and the integration of artificial intelligence and big data.

As industries navigate the complexities of financing, the synergy between commercial banks and businesses remains central to sustainable economic growth. The dynamic nature of banking and credit requires continual adaptation to technological advancements, regulatory changes, and economic shifts, ensuring a resilient and responsive financial ecosystem that fosters industrial development.

Suggested Questions with Solutions

Question 1: Explain the role of commercial banks in industrial development and the economy. How do they facilitate the growth of industries through financial services?

Solution: Commercial banks play a pivotal role in industrial development by providing various financial services. They act as intermediaries between savers and borrowers, channeling funds towards productive activities. Banks offer loans and credit facilities to industries, supporting their capital expenditure, working capital needs, and expansion plans. Additionally, commercial banks provide financial advice, trade finance, and investment banking services, contributing significantly to the overall economic growth.

Question 2: Discuss the importance of credit in industrial functioning. How does the availability of credit impact the performance of industries, especially in developing economies?

Solution: Credit is crucial for industrial functioning as it enables businesses to finance operations, invest in technology, and expand their capacities. In developing economies, where access to capital may be limited, the availability of credit becomes even more critical. Adequate credit allows industries to innovate, improve efficiency, and remain competitive in the global market. Moreover, it promotes entrepreneurship and the establishment of new businesses, fostering economic development and job creation.

Question 3: Enumerate the various sources of finance provided by commercial banks to industries. How do these financial instruments cater to the diverse needs of businesses in different stages of development?

Solution: Commercial banks offer a range of financial instruments to industries, including term loans, working capital loans, overdraft facilities, and letters of credit. These instruments cater to the diverse needs of businesses at various stages of development. Term loans support long-term projects, while working capital loans address short-term operational needs. Overdraft facilities provide flexibility, and letters of credit facilitate international trade. This variety of financial products helps industries manage their finances effectively and adapt to changing market conditions.

Question 4: Assess the risks associated with providing credit to industries. How do commercial banks manage these risks to ensure the stability of their lending portfolios?

Solution: Providing credit to industries involves inherent risks, such as default, market fluctuations, and economic downturns. Commercial banks employ risk management strategies to mitigate these risks. This includes thorough credit assessments, collateral requirements, and periodic monitoring of borrowers. Banks diversify their lending portfolios across industries and regions to reduce concentration risk. Additionally, they use financial derivatives and insurance products to hedge against market volatility, ensuring the stability of their lending activities.

Question 5: Explain the concept of project finance and its relevance in industrial projects. How do commercial banks contribute to project finance, and what benefits does it offer to both banks and industries?

Solution: Project finance involves financing a specific project based on its cash flow and assets, rather than the creditworthiness of the project sponsors. Commercial banks play a crucial role in project finance by structuring financing arrangements tailored to the unique risks and cash flow patterns of each project. This approach allows industries to undertake large and capital-intensive projects without significantly impacting their balance sheets. Project finance benefits both banks and industries by aligning the interests of all stakeholders, enhancing project transparency, and ensuring the efficient allocation of resources.

Question 6: ABC Manufacturing Ltd. approaches XYZ Bank for a term loan of $500,000 to expand its production capacity. The bank offers an interest rate of 8% per annum for a tenure of 5 years. Calculate the total interest payable and the monthly installment using the simple interest method.

Solution:
- Principal Amount (P): $500,000
- Interest Rate (R): 8% per annum
- Time (T): 5 years

Total Interest = P * R * T / 100
= $500,000 * 8 * 5 / 100
= $200,000

Total Repayment Amount = P + Total Interest
= $500,000 + $200,000
= $700,000

Monthly Installment = Total Repayment Amount / (T * 12)
= $700,000 / (5 * 12)
= $11,666.67 (approx)

Question 7: XYZ Company applies for a working capital loan of $1,000,000 from LMN Bank. The bank's credit analysis determines a Debt-Service Coverage Ratio (DSCR) of 1.5. If the company's annual cash flow available for debt service is $750,000, calculate the maximum annual debt service and the loan amount the company can afford.

Solution:
- Cash Flow Available for Debt Service = $750,000
- Debt-Service Coverage Ratio (DSCR) = 1.5

Maximum Annual Debt Service = Cash Flow / DSCR
= $750,000 / 1.5

$$= \$500,000$$

Loan Amount = Maximum Annual Debt Service
$$= \$500,000$$

Question 8: LMN Textiles requires working capital finance for its operations. The current assets are $1,200,000, and current liabilities are $800,000. Calculate the net working capital and comment on the financial health of the company.

Solution:
- Current Assets = $1,200,000
- Current Liabilities = $800,000

Net Working Capital = Current Assets - Current Liabilities
$$= \$1,200,000 - \$800,000$$
$$= \$400,000$$

The positive net working capital indicates that the company has enough short-term assets to cover its short-term liabilities, reflecting a healthy financial position.

C Code for Bank

```c
#include <stdio.h>

// Structure to represent a bank account
struct BankAccount {
    int accountNumber;
    char accountHolder[100];
    float balance;
};

// Function to create a new bank account
struct BankAccount createAccount(int accountNumber, char accountHolder[100]) {
    struct BankAccount newAccount;
    newAccount.accountNumber = accountNumber;
    strcpy(newAccount.accountHolder, accountHolder);
    newAccount.balance = 0.0;
    return newAccount;
}

// Function to deposit money into an account
void deposit(struct BankAccount *account, float amount) {
    account->balance += amount;
    printf("Deposited $%.2f into account %d\n", amount, account->accountNumber);
}

// Function to withdraw money from an account
void withdraw(struct BankAccount *account, float amount) {
    if (amount <= account->balance) {
        account->balance -= amount;
        printf("Withdrawn $%.2f from account %d\n", amount, account->accountNumber);
    } else {
        printf("Insufficient funds in account %d\n", account-
```

```
>accountNumber);
    }
}

// Function to check the balance of an account
void checkBalance(struct BankAccount account) {
    printf("Account        %d        balance:        $%.2f\n",
account.accountNumber, account.balance);
}

int main() {
    // Example usage
    struct BankAccount account1 = createAccount(1, "John Doe");
    deposit(&account1, 1000.0);
    withdraw(&account1, 500.0);
    checkBalance(account1);

    return 0;
}
```

JAVA Code for Bank

```java
import java.util.Scanner;

class BankAccount {
    private int accountNumber;
    private String accountHolder;
    private double balance;

    // Constructor to create a new account
    public BankAccount(int accountNumber, String accountHolder)
{
        this.accountNumber = accountNumber;
        this.accountHolder = accountHolder;
        this.balance = 0.0;
    }

    // Method to deposit money into the account
    public void deposit(double amount) {
        balance += amount;
        System.out.printf("Deposited $%.2f into account %d%n",
amount, accountNumber);
    }

    // Method to withdraw money from the account
    public void withdraw(double amount) {
        if (amount <= balance) {
            balance -= amount;
            System.out.printf("Withdrawn    $%.2f    from    account
%d%n", amount, accountNumber);
        } else {
            System.out.println("Insufficient funds");
        }
    }

    // Method to check the balance of the account
```

```java
    public void checkBalance() {
        System.out.printf("Account      %d      balance:      $%.2f%n",
accountNumber, balance);
    }
}

public class Bank {
    public static void main(String[] args) {
        Scanner scanner = new Scanner(System.in);

        // Example usage
        BankAccount account1 = new BankAccount(1, "John Doe");

        System.out.println("1. Deposit");
        System.out.println("2. Withdraw");
        System.out.println("3. Check Balance");
        System.out.println("4. Exit");

        int choice;
        do {
            System.out.print("Enter your choice: ");
            choice = scanner.nextInt();

            switch (choice) {
                case 1:
                    System.out.print("Enter deposit amount: $");
                    double depositAmount = scanner.nextDouble();
                    account1.deposit(depositAmount);
                    break;
                case 2:
                    System.out.print("Enter withdrawal amount: $");
                    double withdrawalAmount = scanner.nextDouble();
                    account1.withdraw(withdrawalAmount);
                    break;
                case 3:
                    account1.checkBalance();
                    break;
                case 4:
                    System.out.println("Exiting the program. Goodbye!");
```

```
                break;
            default:
                System.out.println("Invalid choice. Please enter a valid
option.");
            }

        } while (choice != 4);

        scanner.close();
    }
}
```

PYTHON Code for Bank

```python
class BankAccount:
    def __init__(self, account_number, account_holder):
        self.account_number = account_number
        self.account_holder = account_holder
        self.balance = 0.0

    def deposit(self, amount):
        self.balance += amount
        print(f"Deposited     ${amount:.2f}     into     account
{self.account_number}")

    def withdraw(self, amount):
        if amount <= self.balance:
            self.balance -= amount
            print(f"Withdrawn     ${amount:.2f}     from     account
{self.account_number}")
        else:
            print("Insufficient funds")

    def check_balance(self):
        print(f"Account          {self.account_number}          balance:
${self.balance:.2f}")


# Example usage
account1 = BankAccount(1, "John Doe")

while True:
    print("1. Deposit")
    print("2. Withdraw")
    print("3. Check Balance")
    print("4. Exit")

    choice = input("Enter your choice: ")
```

```python
if choice == "1":
    deposit_amount = float(input("Enter deposit amount: $"))
    account1.deposit(deposit_amount)
elif choice == "2":
    withdrawal_amount = float(input("Enter withdrawal amount: $"))

    account1.withdraw(withdrawal_amount)
elif choice == "3":
    account1.check_balance()
elif choice == "4":
    print("Exiting the program. Goodbye!")
    break
else:
    print("Invalid choice. Please enter a valid option.")
```

9 RESERVE BANK OF INDIA (RBI) AND ITS FUNCTIONS: A COMPREHENSIVE OVERVIEW

The Reserve Bank of India (RBI) stands as the central banking institution of the Republic of India, wielding significant influence over the country's monetary policy and financial stability. Established on April 1, 1935, under the Reserve Bank of India Act, the RBI has evolved over the years to become a pivotal force in shaping India's economic landscape. This comprehensive exploration will delve into the origins, structure, and multifaceted functions of the RBI, highlighting its critical role in fostering economic growth and maintaining financial stability.

9.1 Historical Background

The genesis of the Reserve Bank of India can be traced back to the early 20th century, amid concerns about the lack of a centralized banking system in the country. The Royal Commission on Indian

Currency and Finance, also known as the Hilton-Young Commission, recommended the establishment of a central bank in India. Consequently, the Reserve Bank of India Act was enacted in 1934, and the RBI commenced its operations on April 1, 1935. Sir Osborne Smith became its first Governor, and the bank began functioning from its headquarters in Kolkata.

9.2 Structure and Organization

The organizational structure of the Reserve Bank of India is designed to ensure efficient governance and effective execution of its diverse functions. The bank operates under the guidance of a central board, headed by the Governor. The central board consists of government-appointed directors, including the Governor, Deputy Governors, and other officials. The bank's management structure also comprises four Deputy Governors who oversee specific areas such as monetary policy, banking supervision, and financial stability. The RBI has its headquarters in Mumbai, and it operates through a network of regional offices and branches located across major cities in India. The decentralized structure enables the RBI to cater to the diverse economic needs of different regions while maintaining a cohesive national monetary policy.

9.3 Monetary Policy

One of the primary functions of the RBI is the formulation and implementation of monetary policy in India. The Monetary Policy Committee (MPC), established in 2016, plays a crucial role in this process. The MPC is responsible for setting the benchmark interest rate (repo rate) to achieve the targeted inflation rate. The RBI adopts an inflation targeting framework, with the Consumer Price Index (CPI) serving as the key metric for determining the inflation target. Through open market operations, repo rates, and other monetary tools, the RBI aims to control inflation, foster economic growth, and maintain price stability. The bank's ability to influence interest rates has a direct impact on borrowing costs, investment decisions, and overall economic activity in the country.

9.4 Currency Management

The Reserve Bank of India holds the exclusive right to issue and manage the currency in the country. It plays a pivotal role in ensuring an adequate supply of currency notes and coins to meet the demands of the economy. The RBI is also responsible for the withdrawal and demonetization of currency notes to combat issues such as counterfeiting and the circulation of black money. The design, production, and distribution of currency notes are meticulously overseen by the RBI. Security features are continually upgraded to prevent counterfeiting, reflecting the bank's commitment to maintaining the integrity of the Indian currency.

9.5 Banking Regulation and Supervision

Another critical function of the RBI is the regulation and supervision of banks and financial institutions operating in India. The central bank formulates policies to ensure the stability and resilience of the banking sector. It issues licenses to new banks, regulates the activities of existing banks, and monitors their compliance with prudential norms. The RBI's role as a banking regulator extends beyond traditional banks to include non-banking financial institutions (NBFCs) and other financial intermediaries. The aim is to safeguard the interests of depositors, maintain financial stability, and prevent systemic risks that could jeopardize the overall health of the financial system.

9.6 Developmental Functions

In addition to its regulatory and monetary policy functions, the RBI actively engages in developmental activities aimed at fostering financial inclusion and economic development. The bank promotes initiatives that enhance the accessibility of banking services to all segments of society, particularly in rural and underserved areas. The RBI's commitment to financial inclusion is evident through measures such as the Pradhan Mantri Jan Dhan Yojana (PMJDY), which seeks to provide financial services, including banking and insurance, to the unbanked population. The bank also encourages the adoption of technology in banking to improve efficiency and widen the reach of financial services.

9.7 Foreign Exchange Management

As the issuer of the Indian Rupee, the RBI is intricately involved in managing the country's foreign exchange reserves. It formulates and implements policies to ensure the stability of the external value of the rupee and facilitates foreign trade and investment. The RBI intervenes in the foreign exchange market to prevent excessive volatility and maintain a balanced and stable exchange rate. The bank's role in foreign exchange management extends to regulating and supervising the foreign exchange market, authorizing foreign exchange transactions, and implementing measures to safeguard the external sector of the economy.

9.8 Financial Stability

The Reserve Bank of India places a strong emphasis on maintaining overall financial stability in the country. This involves monitoring and addressing risks that could undermine the soundness of the financial system. The RBI conducts regular assessments of the financial health of banks and financial institutions, implements stress testing, and devises strategies to mitigate systemic risks. In times of financial distress, the RBI has the authority to intervene and take corrective measures to stabilize the financial system. This may include restructuring troubled financial institutions, injecting liquidity into the market, or implementing regulatory changes to enhance the resilience of the financial sector.

9.9 Crisis Management

The RBI plays a pivotal role in managing financial crises and ensuring the orderly functioning of financial markets during turbulent times. The bank's experience and expertise are particularly evident during economic downturns, global financial crises, or other extraordinary events that pose challenges to the stability of the financial system. The RBI's crisis management efforts may involve liquidity support to banks, regulatory interventions, and coordination with other regulatory bodies and government agencies. The goal is to instill confidence in the financial system and facilitate a prompt recovery.

9.10 Conclusion

In conclusion, the Reserve Bank of India stands as a cornerstone of India's economic framework, wielding extensive powers to shape monetary policy, regulate the financial sector, and foster economic development. Through its multifaceted functions, the RBI plays a crucial role in maintaining price stability, ensuring the integrity of the currency, and safeguarding the overall health of the financial system.

As India continues its journey towards economic growth and financial inclusion, the Reserve Bank of India remains at the forefront, adapting its policies and strategies to meet the evolving needs of a dynamic and diverse economy. The bank's commitment to stability, inclusivity, and innovation positions it as a key architect of India's economic future.

Suggested Questions with Solutions

Question 1: Explain the historical background of the Reserve Bank of India and its significance in the Indian financial system.

Solution: The Reserve Bank of India (RBI) was established on April 1, 1935, under the Reserve Bank of India Act. Its creation was prompted by the recommendations of the Hilton-Young Commission, which highlighted the need for a centralized banking system in India. The RBI's primary objective is to regulate the monetary policy, issue and manage the Indian currency, and ensure the stability of the financial system. Over the years, the RBI has played a crucial role in shaping India's economic landscape, contributing significantly to the country's financial stability and growth.

Question 2: Outline the structure and organization of the Reserve Bank of India, including the roles of the central board and the Monetary Policy Committee (MPC).

Solution: The RBI operates under the governance of a central board, headed by the Governor. The central board includes government-appointed directors, such as the Governor, Deputy Governors, and other officials. The central board is responsible for overseeing the general administration and operations of the RBI. Additionally, the Monetary Policy Committee (MPC) was established in 2016 to determine the key policy rates and formulate the monetary policy of the country. The MPC consists of six members, including three from the RBI and three government-appointed external members.

Question 3: Examine the role of the Reserve Bank of India in currency management. How does the RBI contribute to the design, production, and distribution of currency notes in the country?

Solution: The RBI holds the exclusive right to issue and manage the currency in India. It is responsible for designing, printing, and distributing currency notes and coins. The RBI ensures an adequate supply of currency to meet the demands of the economy. The bank also takes measures to enhance the security features of currency notes, preventing counterfeiting and maintaining the integrity of the Indian currency.

Question 4: Discuss the developmental functions of the Reserve Bank of India. Provide examples of initiatives taken by the RBI to promote financial inclusion and economic development.

Solution: The RBI actively engages in developmental activities to foster financial inclusion and economic development. Initiatives such as the Pradhan Mantri Jan Dhan Yojana (PMJDY) exemplify the RBI's commitment to providing financial services to the unbanked population. The bank also encourages the adoption of technology in banking to improve efficiency and widen the reach of financial services. These efforts contribute to the overall development and inclusivity of the financial sector.

Question 5: Explain the role of the Reserve Bank of India in maintaining financial stability. How does the RBI address systemic risks and financial crises in the country?

Solution: The RBI plays a critical role in maintaining overall financial stability by monitoring and addressing risks to the financial system. The bank conducts regular assessments of the financial health of banks and financial institutions, implements stress testing, and takes corrective measures when necessary. In times of financial distress, the RBI intervenes to stabilize the financial system, which may involve liquidity support, regulatory interventions, and coordination with other regulatory bodies and government agencies.

10 BUSINESS MANAGEMENT AND ORGANIZATION: PROPRIETORSHIP, PARTNERSHIP, AND JOINT-STOCK COMPANY

10.1 Introduction

𝔅usinesses are the lifeblood of economies worldwide, driving innovation, creating employment, and contributing to economic growth. The structure and organization of a business play a crucial role in its success, and entrepreneurs must carefully consider various factors before choosing a suitable form. In this comprehensive exploration, we delve into three fundamental types of business entities: proprietorship, partnership, and joint-stock company. We will examine their formation, characteristics, advantages, and disadvantages, providing practical examples from India and abroad to illustrate key concepts.

10.2 Proprietorship

A proprietorship, also known as a sole proprietorship, is a business structure characterized by a single individual owning and managing the entire enterprise. In this form of business ownership, the owner is personally responsible for all aspects of the business, including its profits and losses. The proprietor has complete control over decision-making, allowing for a quick and straightforward decision-making process. However, this structure also implies that the proprietor bears unlimited personal liability for the business's debts and obligations, risking personal assets in case of business-related liabilities. Proprietorships are often small-scale enterprises and are prevalent in businesses like consulting services, retail shops, or freelance work. While offering simplicity and autonomy, the proprietorship structure may face limitations in terms of raising capital and expanding operations compared to other business entities like partnerships or corporations.

10.2.1 Definition and Characteristics

A sole proprietorship is the simplest form of business organization, where a single individual owns and manages the entire enterprise. This individual assumes all responsibilities, risks, and rewards associated with the business.

10.2.2 Formation

The formation of a proprietorship is relatively straightforward. It involves no formal legal processes or extensive documentation. In most jurisdictions, the business owner can start operating by obtaining any required licenses or permits.

10.2.3 Advantages

- **Sole Control:** The proprietor has complete control over decision-making and operations.
- **Ease of Formation:** Quick and easy to establish with minimal bureaucratic processes.
- **Tax Benefits:** Income from the business is typically taxed as personal income, offering potential tax advantages.

10.2.4 Disadvantages

- **Limited Capital:** Limited financial resources as the proprietor's personal funds are the primary source of capital.
- **Unlimited Liability:** The proprietor is personally responsible for

all debts and liabilities, risking personal assets.

- **Limited Expertise:** Relies on the skills and expertise of a single individual, limiting the scope of knowledge and experience.

10.2.5 Practical Example in India: Kirana Stores

In India, countless kirana (mom-and-pop) stores exemplify proprietorship. These small retail businesses, often operated by a single individual or a family, showcase the simplicity and flexibility of this business structure.

10.3 Partnership

Partnership is a form of business organization where two or more individuals come together to manage and operate a business in accordance with the terms and objectives set out in a Partnership Deed. This legal document outlines the rights, responsibilities, profit-sharing arrangements, and decision-making authority of each partner. There are various types of partnerships, including general partnerships where all partners have equal responsibility and liability, and limited partnerships where some partners have limited liability based on their investment. Partnerships benefit from the combined skills, resources, and capital of the individuals involved, fostering collaboration and shared responsibilities. However, one key consideration is that each partner is personally liable for the debts and obligations of the business. Partnerships are common in professional services, legal firms, and small businesses where the expertise and contributions of multiple individuals are essential for success. Clear communication, trust, and a well-drafted partnership agreement are crucial for the smooth functioning of the partnership.

10.3.1 Definition and Characteristics

A partnership involves two or more individuals pooling their resources, skills, and capital to operate a business. Partnerships are governed by a partnership deed, a legal document outlining the terms and conditions of the partnership.

10.3.2 Formation

The formation of a partnership involves drafting and signing a partnership deed. This document typically includes details such as profit-sharing ratios, roles and responsibilities, and dispute resolution mechanisms.

10.3.3 Advantages

- **Shared Responsibility:** Partners share the responsibilities, risks, and decision-making.
- **Additional Capital:** The presence of multiple partners allows for a greater pool of financial resources.
- **Broader Skill Set:** Partnerships often benefit from the diverse skills and expertise of the individuals involved.

10.3.4 Disadvantages

- **Disagreements:** Differences in opinion or conflicts among partners can lead to disputes.
- **Unlimited Liability:** In a general partnership, each partner is personally liable for the business debts.
- **Limited Capital:** While there is access to more capital than in a proprietorship, it may still be limited compared to larger business structures.

10.3.5 Practical Example Abroad: Law Firms

Law firms often operate as partnerships, particularly in countries like the United States and the United Kingdom. Partners in a law firm collaborate, combining legal expertise, financial resources, and client networks to provide comprehensive legal services.

10.4 Joint-Stock Company

A joint-stock company is a form of business organization that issues shares of stock to individuals who become shareholders or stockholders. These shares represent ownership in the company and confer certain rights, such as voting on key decisions and receiving dividends from profits. Joint-stock companies are characterized by shared ownership, allowing for the pooling of capital from multiple investors to finance the company's operations and expansion. The liability of individual shareholders is generally limited to the value of their shares, providing a level of protection for personal assets. Joint-stock companies can be publicly traded on stock exchanges, allowing shares to be bought and sold by the public, or privately held with a limited number of shareholders. This structure enables companies to access significant capital, facilitating large-scale projects and promoting economic development. However, it also involves complex regulatory requirements, transparency expectations, and the challenge of balancing diverse shareholder interests.

10.4.1 Definition and Characteristics

A joint-stock company, also known as a corporation, is a legal entity separate from its owners (shareholders). Ownership is determined by the number of shares held, and the company is managed by a board of directors elected by the shareholders.

10.4.2 Formation

The formation of a joint-stock company is a complex process involving legal formalities. It requires the issuance of shares, the creation of articles of association and memorandum of association, and compliance with regulatory requirements.

10.4.3 Advantages

- **Limited Liability:** Shareholders' liability is limited to the amount invested in the company.
- **Enhanced Capital:** The company can raise significant capital by issuing shares to the public.
- **Perpetual Existence:** The company exists independently of its shareholders, ensuring continuity even if shareholders change.

10.4.4 Disadvantages

- **Complexity and Cost:** Formation involves complex legal processes and can be costly.
- **Regulatory Compliance:** Strict regulatory requirements must be adhered to, adding to administrative burdens.
- **Loss of Control:** Shareholders may have limited control over the day-to-day operations, with decisions often made by the board of directors.

10.4.5 Practical Example in India: Infosys

Infosys, one of India's leading IT companies, is an example of a joint-stock company. It went public in 1993, issuing shares to the public to raise capital. The company is governed by a board of directors, and its shareholders enjoy limited liability.

10.5 Comparative Analysis

Comparative analysis is a systematic approach that involves evaluating and comparing two or more entities, systems, processes, or phenomena to discern similarities, differences, and patterns. This analytical method is widely used in various disciplines, including business, finance, literature, and scientific research. In business, for instance, a comparative analysis may be conducted on the financial performance of competing companies to identify strengths, weaknesses, and areas for improvement. It often involves examining financial ratios, market share, and operational efficiency. Comparative analysis is instrumental in decision-making, strategic planning, and understanding the competitive landscape. It provides valuable insights that help stakeholders make informed choices by highlighting benchmarks, trends, and best practices. Whether applied to financial statements, product features, or performance metrics, comparative analysis serves as a powerful tool for gaining a deeper understanding of complex systems and making informed, data-driven decisions.

10.5.1 Flexibility

- **Proprietorship:** Offers maximum flexibility as decisions are made by a single individual.
- **Partnership:** Allows for shared decision-making but may still face challenges in reaching a consensus.
- **Joint-Stock Company:** Decision-making is typically in the hands of the board of directors, with limited involvement from shareholders.

10.5.2 Liability

- **Proprietorship:** Unlimited personal liability for the proprietor.
- **Partnership:** Partners share unlimited liability, but limited partners in a limited partnership have liability restricted to their investment.
- **Joint-Stock Company:** Shareholders enjoy limited liability.

10.5.3 Capital

- **Proprietorship:** Limited to the personal funds of the proprietor.
- **Partnership:** Capital is increased by contributions from multiple partners.

- **Joint-Stock Company:** Can raise significant capital by issuing shares to the public.

10.5.4 Decision-Making

- **Proprietorship:** Centralized decision-making by the proprietor.
- **Partnership:** Shared decision-making, but potential for conflicts.
- **Joint-Stock Company:** Decisions are made by the board of directors, with limited input from shareholders.

10.6 Conclusion

Choosing the right business structure is a critical decision for entrepreneurs, shaping the trajectory of their ventures. Proprietorship, partnership, and joint-stock companies each have unique characteristics, advantages, and disadvantages. The choice depends on factors such as the scale of operations, risk tolerance, and desired level of control.

In India and abroad, diverse examples showcase the versatility of these business structures. From the simplicity of kirana stores to the complexity of multinational corporations like Infosys, the business landscape reflects the adaptability of these organizational forms.

Understanding the nuances of proprietorship, partnership, and joint-stock companies empowers entrepreneurs to make informed decisions that align with their vision and goals. As the global business environment continues to evolve, the significance of these organizational structures remains paramount in shaping the future of commerce and industry.

Suggested Questions with Solutions

1. Question: Explain the advantages and disadvantages of a sole proprietorship. Provide examples to illustrate each point.

Solution:
Advantages of Sole Proprietorship:
- Full control and decision-making authority.
- Direct and simple taxation.
- Easy and inexpensive to establish.

Disadvantages of Sole Proprietorship:
- Limited resources and capital.
- Unlimited personal liability.
- Lack of continuity and succession planning.

2. Question: Compare and contrast general partnerships and limited partnerships. Provide a scenario where each type would be most suitable.

Solution:
General Partnership:
- All partners have equal management authority and share both profits and liabilities.
- Suitable for small businesses where partners have similar roles and responsibilities.

Limited Partnership:
- Involves both general and limited partners, where limited partners have restricted liability.
- Suitable when investors want to contribute capital but avoid active management roles.

3. Question: Discuss the key features and benefits of a joint-stock company. Provide an example of a successful joint-stock

company and its impact on the market.

Solution:
Key Features of Joint-Stock Company:
- Limited liability for shareholders.
- Transferability of shares.
- Perpetual existence.

Example:
- Apple Inc. is a successful joint-stock company known for its innovation and market dominance, with shares traded on stock exchanges.

4. Question: Outline the steps involved in the formation of a partnership. What legal agreements are essential, and how do they protect the interests of each partner?

Solution:
Steps in Formation:
1. Mutual agreement among partners.
2. Drafting a partnership deed specifying roles, responsibilities, and profit-sharing.
3. Registration of the partnership with relevant authorities.

Legal Agreements:
- Partnership Deed: Outlines terms and conditions.
- Articles of Partnership: Defines internal rules and regulations.

5. Question: Analyze the factors that influence the choice between a private and a public joint-stock company. Provide real-world examples of companies that have transitioned from private to public.

Solution:
Factors Influencing Choice:
- Capital requirements.

- Reporting and disclosure obligations.
- Control and decision-making preferences.

Examples:

- Facebook (now Meta) transitioned from private to public in 2012, allowing for increased capital through an Initial Public Offering (IPO).

11 EXPLORING THE SYNERGY OF FINANCE AND MANAGEMENT: AN IN-DEPTH ANALYSIS OF TAXATION, INSURANCE, BUSINESS COMBINATIONS, AND BASIC PRINCIPLES OF MANAGEMENT

11.1 Introduction

In the complex and dynamic landscape of business, the interplay between finance and management is crucial for organizational success. This comprehensive analysis delves into key elements of finance and management, exploring taxation, insurance, business combinations, and basic principles of management. Each element plays a pivotal role in shaping the financial health and managerial efficiency of an organization. This multidimensional exploration delves into the nuanced aspects of taxation, shedding light on its impact on financial strategies and managerial decisions. Simultaneously, the investigation extends to the realm of insurance, elucidating its role in risk mitigation and financial planning. Moreover, the study navigates through the intricacies of business combinations, unraveling the intricate financial and managerial tapestry woven in mergers, acquisitions, and partnerships. Anchored in fundamental principles, the discourse unearths the bedrock of effective management, highlighting its synergistic dance with financial strategies for sustainable organizational growth. This analysis promises a profound understanding of how the amalgamation of finance and management fosters a resilient framework for navigating the complexities of the contemporary business landscape.

11.2 Elements of Taxation

Elements of taxation encompass a multifaceted framework that underpins the financial operations of governments and organizations alike. At its core, taxation involves key components that structure the imposition and collection of levies to fund public services and societal needs. One fundamental element is the tax base, representing the economic value subject to taxation, whether it be income, consumption, or property. Tax rates constitute another crucial facet, delineating the percentage at which the tax base is assessed. The design of tax systems also involves considerations of equity, efficiency, and simplicity, reflecting the overarching principles guiding tax policy. Additionally, tax administration plays a pivotal role, encompassing the processes of collection, enforcement, and compliance monitoring. Understanding the elements of taxation involves a nuanced exploration of these components, recognizing their intricate interplay in shaping fiscal policies and contributing to the financial stability of governments and organizations worldwide.

11.2.1 Taxation Overview

Taxation is a critical component of financial management, influencing business decisions and shaping the economic landscape. Understanding the various elements of taxation is essential for effective financial planning and compliance.

11.2.2 Types of Taxes

- **Income Tax:** Income tax, a ubiquitous feature of fiscal systems globally, manifests differently in the tax landscapes of India and other foreign countries, reflecting unique socio-economic considerations and legislative frameworks. In India, the income tax system is progressive, with tax rates varying based on income slabs. The country employs a residency-based taxation system, taxing residents on their global income, while non-residents are taxed only on income generated within India. Deductions and exemptions are integral elements, offering taxpayers avenues to reduce their taxable income. Contrastingly, foreign countries often adopt diverse models, such as the flat tax system or a combination of progressive and regressive elements. The principles of residency and source play a crucial role, determining the extent to which global income is subject to taxation. Tax treaties between countries aim to mitigate double taxation concerns. While India has made strides in digital taxation, the global landscape grapples with evolving challenges, such as base erosion and profit shifting. The comparative study of income tax in India and foreign jurisdictions underscores the intricate nuances that shape fiscal policies and international tax relations in an increasingly interconnected world.

- **Corporate Tax:** Corporate tax is a form of taxation imposed on the profits earned by businesses and corporations. This tax is typically levied by government authorities on the net income generated by a company during a specific period. The primary objective of corporate tax is to generate revenue for the government and contribute to public services and infrastructure development. The tax rate varies across countries and often depends on the size and profitability of the corporation. Governments may offer incentives or deductions to encourage certain types of business activities, stimulate economic growth, or attract foreign investments. Corporate tax plays a crucial role in the overall fiscal policy of a nation, impacting the competitiveness of businesses and influencing investment decisions. Policymakers continually evaluate and adjust corporate tax policies to strike a balance between fostering economic development and ensuring a fair distribution of the tax burden.

- **Sales Tax:** Sales tax is a consumption-based tax imposed by governments on the sale of goods and services at the retail level. Unlike income or corporate taxes, sales tax is applied directly to the final purchase price paid by the consumer. The tax rate varies across jurisdictions, and it may be a fixed percentage of the sale price or vary based on the type of goods or services. Sales tax is a crucial revenue source for governments, funding public services and infrastructure. Businesses acting as intermediaries in the supply chain are responsible for collecting and remitting the sales tax to the government. The regressive nature of sales tax means that it affects all consumers regardless of income, although some essential items may be exempt or subject to reduced rates to alleviate the burden on lower-income individuals. Policymakers often navigate the balance between generating revenue and ensuring the tax system's fairness and efficiency when designing and implementing sales tax policies.

11.2.3 Tax Planning and Management

Tax planning and management encompass strategic financial practices aimed at optimizing an individual or business entity's tax liability within the legal framework. This proactive approach involves careful analysis of financial activities, investments, and transactions to minimize the overall tax burden. For individuals, tax planning may involve decisions related to income, deductions, investments, and retirement contributions. Businesses engage in tax planning to optimize their corporate structure, take advantage of available tax credits and incentives, and ensure compliance with tax regulations. Efficient tax management involves timely filing, accurate record-keeping, and compliance with ever-changing tax laws. Skilled tax planning can result in significant cost savings, increased cash flow, and improved financial efficiency. However, it's crucial to strike a balance between legitimate tax planning and ethical considerations, avoiding aggressive tax avoidance schemes that may attract legal scrutiny. Professional advice from tax experts is often sought to navigate the complexities of tax planning and management effectively.

- **Tax Optimization:** Tax optimization refers to the strategic and legal methods employed by individuals and businesses to minimize their tax liabilities while staying within the bounds of existing tax laws. This process involves analyzing financial activities, investments, and transactions to identify opportunities for reducing taxable income, maximizing deductions, and taking advantage of available tax credits and incentives. For businesses, tax optimization often includes structuring operations efficiently, exploring tax-efficient investment strategies, and utilizing available tax breaks. Individuals may engage in tax optimization through retirement planning, charitable contributions, and investment decisions. While tax optimization is a legitimate and common practice, it is important to distinguish it from illegal tax evasion. The goal of tax optimization is to enhance financial efficiency and ensure that taxpayers pay the appropriate amount of taxes in a lawful and ethical manner. Consulting with tax professionals or financial advisors is common for those seeking to navigate the complexities of tax optimization successfully.

- **Compliance:** Compliance is a cornerstone of effective tax planning and management, emphasizing adherence to relevant tax laws, regulations, and reporting requirements. In the realm of taxation, compliance ensures that individuals and businesses fulfill their legal obligations, file accurate and timely returns, and transparently report financial activities. Successful tax planning and management require a thorough understanding of the intricate and ever-evolving tax codes, both at the national and regional levels. Failing to comply with tax regulations can lead to severe consequences, including penalties, fines, and legal actions. Therefore, a comprehensive approach to tax planning involves not only optimizing financial strategies but also prioritizing meticulous record-keeping and a commitment to ethical practices. Many businesses and individuals enlist the expertise of tax professionals to navigate the complexities of tax compliance, ensuring that their financial activities align with the prevailing legal requirements while still benefiting from strategic tax planning measures.

11.2.4 International Taxation

International taxation is a complex field that deals with the taxation of cross-border economic activities involving individuals, businesses, and investments across different countries. The fundamental challenge lies in coordinating tax policies and resolving potential conflicts among various jurisdictions. Countries may tax income based on residency, source, or a combination of both, leading to intricate issues of double taxation or opportunities for tax planning. International tax laws often address the allocation of taxing rights and establish mechanisms, such as tax treaties, to mitigate double taxation and prevent tax evasion. Multinational corporations engage in sophisticated tax planning to optimize their global tax liabilities, taking advantage of variations in tax rates and incentives among countries. As the global economy becomes more interconnected, international taxation has gained prominence, prompting ongoing efforts by governments and international organizations to create frameworks that promote fairness, transparency, and cooperation in the realm of cross-border taxation.

11.2.5 Transfer Pricing

Transfer pricing is a critical aspect of international taxation that deals with the pricing of transactions between affiliated companies or entities within a multinational corporation. It involves determining the prices of goods, services, or intellectual property transferred between related entities, and these prices can significantly impact the allocation of profits and tax liabilities among different jurisdictions. The objective of transfer pricing regulations is to ensure that transactions between related parties are conducted at arm's length—meaning the prices are set as if the entities were unrelated—to prevent the manipulation of profits for tax purposes. Governments implement transfer pricing rules to maintain tax fairness, prevent profit shifting, and safeguard their tax bases. Multinational corporations must adhere to these regulations, and compliance often requires careful documentation, benchmarking studies, and alignment with the arm's length principle. Effective transfer pricing practices contribute to a transparent and equitable international tax system while helping to prevent tax avoidance and evasion.

11.2.6 Double Taxation

Double taxation occurs when a taxpayer is subject to taxation on the same income in two or more different jurisdictions. This can arise when a person or a business earns income that is taxable in both the country where it is generated and the country of residence. The most common forms of double taxation are economic double taxation and juridical double taxation.

Economic double taxation refers to the taxation of the same income by two or more countries without any relief or credit for taxes paid in the other jurisdiction. This situation can lead to a significant tax burden on the taxpayer, reducing the incentive for cross-border investments and economic activities.

Juridical double taxation, on the other hand, occurs when two or more countries claim the right to tax the same income based on their respective domestic laws. To address this issue, many countries enter into double tax treaties (DTTs) or agreements to mitigate the impact of juridical double taxation. These treaties typically allocate taxing rights between the contracting states, provide mechanisms for the relief of double taxation, and establish procedures for the exchange of information between tax authorities.

To alleviate the burden of double taxation, some countries adopt unilateral measures, such as allowing a tax credit for foreign taxes paid or providing exemptions for certain types of income. International efforts, including those by organizations like the Organisation for Economic Co-operation and Development (OECD), aim to develop common standards and guidelines to address issues related to double taxation, promoting fairness and efficiency in the global tax system.

11.3 Insurance in Financial Management

Insurance plays a crucial role in financial management by providing a mechanism to mitigate risks and uncertainties. In the realm of personal finance, individuals often purchase various insurance policies to protect themselves and their families from financial setbacks caused by unexpected events. Common types of personal insurance include life insurance, health insurance, property insurance, and disability insurance. Life insurance, for example, ensures financial security for the policyholder's beneficiaries in the event of the policyholder's death.

Similarly, businesses incorporate insurance into their financial management strategies to safeguard against potential losses and liabilities. Commercial insurance covers a range of risks, including property damage, liability claims, business interruption, and employee-related risks. By transferring these risks to an insurance company through premium payments, businesses can enhance their financial resilience and maintain operational continuity in the face of unforeseen events.

Risk management is a fundamental aspect of financial management, and insurance serves as a key tool in this process. It allows individuals and businesses to allocate the financial impact of risks to an external entity, providing a sense of security and stability in an unpredictable environment. Moreover, insurance enables financial planning by helping individuals and businesses estimate and budget for potential future losses, fostering a more systematic and strategic approach to financial management. In summary, insurance is an integral component of financial planning and risk management, providing a safety net for individuals and businesses alike.

11.3.1 Role of Insurance

The role of insurance is multifaceted, serving as a critical component in both personal and business financial management. At its core, insurance acts as a risk management tool, providing protection against various uncertainties that could lead to financial losses. In the realm of personal finance, individuals invest in insurance to safeguard themselves and their families from unforeseen events. Life insurance, for instance, ensures that beneficiaries receive financial support in the event of the policyholder's death, offering peace of mind and financial security.

For businesses, insurance plays a pivotal role in protecting against the myriad risks associated with operations. Property insurance guards against losses due to damage or destruction of physical assets, liability insurance shields against legal claims and lawsuits, and business interruption insurance helps mitigate the financial impact of unexpected disruptions to normal operations. This risk transfer mechanism allows businesses to focus on their core activities without being unduly burdened by the potential financial fallout from unexpected events.

Insurance also facilitates economic stability and growth by encouraging investment and entrepreneurship. Lenders often require businesses to have insurance coverage, reducing the financial risks associated with loans and fostering a more conducive environment for investment. Additionally, insurance can act as a catalyst for innovation and economic development by providing a safety net for individuals and businesses to take calculated risks and pursue opportunities that might otherwise seem too uncertain.

In the broader context, insurance contributes to the overall resilience of societies by promoting financial security and stability. It provides individuals and businesses with the confidence to navigate an ever-changing and unpredictable world. Moreover, insurance fosters a sense of social responsibility, as policyholders collectively contribute to a pool of funds that can be utilized to support those facing adversity. In summary, the role of insurance extends far beyond financial protection, encompassing risk management, economic stability, and the facilitation of growth and innovation in both personal and business contexts.

11.3.2 Types of Insurance

There are various types of insurance designed to address different aspects of risk and uncertainty in both personal and business spheres. Life insurance is a fundamental type that provides financial protection to beneficiaries in the event of the policyholder's death, offering various forms such as term life and whole life insurance. Health insurance covers medical expenses, ensuring individuals have access to necessary healthcare without facing exorbitant costs. Property and casualty insurance, including homeowners, renters, and auto insurance, protect against losses related to physical assets and liabilities. Disability insurance provides income replacement in case of an individual's inability to work due to a disability. Business insurance categories include property insurance, liability insurance, and business interruption insurance, each addressing specific risks faced by enterprises. Additionally, specialized insurance exists, such as travel insurance, which covers unexpected events during trips, and pet insurance, which helps manage veterinary expenses. The diversity of insurance types reflects the broad spectrum of risks individuals and businesses encounter, allowing for tailored risk management solutions in various aspects of life and commerce.

11.3.3 Risk Assessment and Mitigation

Risk assessment and mitigation are integral components of effective risk management strategies, crucial for individuals, businesses, and organizations to navigate uncertainties and potential challenges. Risk assessment involves the identification, analysis, and evaluation of potential risks that could impact objectives or outcomes. This process aims to understand the nature and severity of risks, assess their likelihood of occurrence, and prioritize them based on their potential impact.

Once risks are identified and assessed, the next step is mitigation, which involves developing strategies and actions to reduce or eliminate the impact and likelihood of these risks. Mitigation strategies vary depending on the nature of the risks but often include a combination of risk avoidance, risk reduction, risk transfer, and risk acceptance.

Risk avoidance involves steering clear of activities or situations that pose a significant risk. Risk reduction strategies focus on implementing measures to lessen the severity or likelihood of a risk. This may involve implementing safety protocols, redundancies, or improved processes. Risk transfer involves shifting the impact of the risk to another party, often through insurance or outsourcing. Risk acceptance acknowledges that certain risks are inherent and unavoidable, and organizations choose to absorb the potential impact rather than allocate resources to mitigate them.

In both personal and business contexts, effective risk assessment and mitigation contribute to better decision-making, increased resilience, and the ability to proactively address challenges. It requires a comprehensive understanding of the internal and external factors that may pose a threat, ongoing monitoring and reassessment of risks, and the flexibility to adapt mitigation strategies as circumstances change. By incorporating risk management practices into overall planning and decision-making processes, individuals and organizations can enhance their ability to achieve objectives while navigating an ever-changing and uncertain environment.

11.3.4 Business Interruption Insurance

Business Interruption Insurance (BI) is a type of insurance coverage designed to protect businesses from financial losses incurred during periods of interruption or suspension of normal operations. This interruption could result from various perils, including natural disasters, fires, or other covered events that cause physical damage to the insured property. While property insurance typically covers the physical damage to buildings and equipment, BI steps in to address the consequential financial losses that occur due to the business's inability to operate.

BI insurance typically covers expenses such as lost revenue, ongoing fixed costs like rent and utilities, and even extra expenses incurred to mitigate the impact of the interruption. The coverage may also extend to cover the costs of temporary relocation, if necessary, and expenses related to the restoration of operations.

The calculation of the coverage often involves a detailed assessment of the business's historical financial records and projections. The goal is to ensure that the coverage is adequate to sustain the business during the interruption period, allowing it to recover and resume normal operations. The importance of BI Insurance became particularly evident during events like the COVID-19 pandemic, where many businesses faced unexpected and prolonged disruptions. However, it's essential for businesses to carefully review policy terms and exclusions to understand the scope of coverage fully. BI insurance is a critical component of risk management for businesses, providing financial support during challenging times and helping to ensure continuity in the face of unforeseen events.

11.3.5 Coverage Scope

The coverage scope of insurance refers to the extent of protection and the specific risks that an insurance policy addresses. Different types of insurance policies have distinct coverage scopes tailored to the nature of the risks they are designed to mitigate. For example, in health insurance, coverage may include medical expenses, hospitalization, prescription medications, and preventive care. Life insurance, on the other hand, covers the financial impact of a policyholder's death by providing a death benefit to beneficiaries.

Property and casualty insurance, which includes homeowners and auto insurance, offers coverage for physical assets and liabilities. Homeowners insurance may cover damage to the home, personal belongings, and liability for injuries that occur on the property. Auto insurance typically covers damage to the insured vehicle, liability for injuries or property damage caused by the insured vehicle, and sometimes additional coverage like comprehensive or collision.

Business insurance encompasses a wide range of coverage options, such as property insurance for physical assets, liability insurance to protect against legal claims, and business interruption insurance to address financial losses during operational interruptions.

The coverage scope is outlined in the insurance policy, including specific details about what is covered, any exclusions, limitations, and the terms and conditions of the coverage. Understanding the coverage scope is crucial for policyholders to ensure they have adequate protection for their specific needs and to avoid any surprises in the event of a claim.

Insurance policies often allow for customization through endorsements or riders, enabling policyholders to tailor coverage to their unique circumstances. Regular reviews and updates to insurance coverage are essential to ensure that it remains aligned with changing needs and circumstances. Clear communication with insurance providers and a comprehensive understanding of the coverage scope empower individuals and businesses to make informed decisions about their risk management strategies.

11.3.6 Contingency Planning

Contingency planning is a crucial aspect of both tax planning and insurance, as it involves preparing for unexpected events and uncertainties that may impact financial stability. In tax planning, individuals and businesses engage in contingency planning by considering potential changes in tax laws, economic conditions, or personal financial situations. This involves developing strategies to adapt to these changes, ensuring that tax planning remains effective and aligned with current circumstances. For businesses, contingency planning in tax matters may include scenario analysis, budget adjustments, and proactive measures to capitalize on available tax incentives or credits.

Insurance also relies heavily on contingency planning to address unforeseen events that may lead to financial losses. Businesses and individuals evaluate potential risks, such as natural disasters, accidents, or legal liabilities, and then develop contingency plans that may involve insurance coverage. This ensures that, in the event of a covered loss, the financial impact is mitigated, and the insured party can recover more quickly.

The synergy between tax planning and insurance in contingency planning is evident in risk management strategies. For instance, a business might strategically choose insurance coverage that aligns with potential tax advantages. Certain insurance premiums may be tax-deductible, offering a tax planning opportunity while simultaneously providing financial protection.

In times of economic uncertainty or major life changes, individuals may revisit both their tax planning and insurance strategies as part of a comprehensive contingency plan. Adjusting investment portfolios, reassessing insurance coverage, and optimizing tax positions are common elements of such planning.

Overall, effective contingency planning involves a holistic approach that integrates tax planning and insurance strategies. By anticipating potential challenges and developing proactive responses, individuals and businesses can enhance their resilience and navigate uncertainties with greater confidence and financial security. Regular reviews and updates to contingency plans ensure they remain relevant and responsive to evolving circumstances in both the tax and insurance landscapes.

11.4 Business Combinations

Business combinations refer to transactions where two or more separate entities combine their operations to form a single economic entity. These transactions can take various forms, such as mergers, acquisitions, consolidations, or joint ventures. The primary objective of business combinations is often to achieve synergies that enhance the overall value and competitiveness of the combined entity.

Mergers occur when two or more companies agree to merge into a single entity, with the shareholders of each participating company becoming shareholders of the new entity. Acquisitions, on the other hand, involve one company acquiring the assets, liabilities, and often the controlling interest in another company. Consolidations occur when two or more companies combine to form an entirely new entity, and joint ventures involve the creation of a new entity by two or more parties, each contributing assets and sharing control.

Business combinations can result in various synergies, such as cost savings, increased market share, improved operational efficiency, and access to new technologies or markets. However, these transactions also pose challenges, including cultural integration, regulatory compliance, and financial reporting complexities.

From a financial reporting perspective, business combinations are accounted for using the acquisition method, where the acquiring company recognizes the fair values of the acquired assets and liabilities. Goodwill, representing the excess of the purchase price over the fair value of identifiable assets acquired, is also recognized.

Business combinations play a significant role in corporate strategy, allowing companies to strengthen their market positions, diversify their product or service offerings, and achieve growth objectives. The success of a business combination often depends on effective integration planning and execution, as well as thorough due diligence to assess the potential risks and benefits of the transaction.

11.4.1 Mergers and Acquisitions (M&A)
Mergers and acquisitions (M&A) are strategic business activities involving the combination of two or more companies to create a single entity or the acquisition of one company by another. These transactions are driven by various objectives, including achieving economies of scale, expanding market presence, gaining access to new technologies or talent, and enhancing competitiveness. Mergers and acquisitions can take different forms, such as horizontal mergers involving companies in the same industry, vertical mergers involving companies in different stages of the supply chain, and conglomerate mergers between unrelated businesses.

In a merger, two companies agree to combine their operations, assets, and liabilities to form a new entity. This new entity often represents a blending of the strengths of both merging companies. On the other hand, acquisitions involve one company purchasing the assets or shares of another company. The acquiring company gains control over the target company and assumes its operations.

Mergers and acquisitions are complex transactions that require careful planning, due diligence, and negotiation. Factors such as financial health, market position, management capabilities, and cultural compatibility are critical considerations in the success of these deals. M&A activities can create synergies, allowing the combined entity to realize cost savings, improved efficiency, and increased market power. However, challenges such as integration issues, cultural differences, and regulatory hurdles must be addressed to ensure a smooth transition.

From a financial perspective, M&A transactions involve valuing the target company, negotiating the terms of the deal, and determining the appropriate financing structure. The accounting treatment of M&A transactions follows specific guidelines, with the acquiring company recognizing the fair values of the acquired assets and liabilities, as well as any goodwill arising from the excess of the purchase price over the fair value of identifiable assets.

While M&A activities offer significant strategic advantages, they also carry risks, and not all transactions lead to success. Effective post-merger integration and careful consideration of cultural, operational, and financial factors are key to realizing the intended benefits of mergers and acquisitions.

11.4.2 Mergers

Mergers, which involve the combination of two or more companies to form a single entity, are driven by various strategic and operational reasons. These transactions can take different forms, including horizontal mergers, where companies in the same industry merge, vertical mergers involving companies in different stages of the supply chain, and conglomerate mergers between unrelated businesses. The motivations behind mergers are diverse and can significantly impact the competitive landscape of industries. Here are some common reasons for mergers:

1. Synergy Creation: One of the primary reasons for mergers is the pursuit of synergy. Synergy occurs when the combined entity can achieve efficiencies and benefits that are greater than what each company could achieve individually. This can include cost savings, improved operational efficiency, and enhanced revenue opportunities.

2. Market Expansion: Mergers provide companies with an opportunity to expand their market presence, either geographically or within specific market segments. By combining forces, companies can access new customer bases and distribution channels, leading to increased market share and competitiveness.

3. Diversification: Companies often pursue mergers to diversify their product or service offerings. This strategy helps mitigate risks associated with dependence on a single product or market. Diversification through mergers allows companies to navigate economic cycles and changes in consumer preferences more effectively.

4. Economies of Scale: Achieving economies of scale is a common objective in mergers. Larger entities can often benefit from reduced per-unit costs due to increased production volume, shared resources, and improved bargaining power with suppliers.

5. Technology and Innovation: Mergers can be driven by a desire to acquire advanced technologies or innovative capabilities. This is particularly relevant in industries where rapid technological advancements are critical for maintaining competitiveness.

6. Access to Talent: Mergers provide companies with access to a broader talent pool. Acquiring a company with skilled employees, specialized knowledge, or a strong management team can contribute to the overall success of the merged entity.

7. Financial Synergy: Mergers can enhance financial performance through improved access to capital, increased profitability, and strengthened financial stability. This can be particularly attractive to companies facing financial challenges or seeking to optimize their capital structure.

8. Regulatory Compliance: Companies may engage in mergers to ensure compliance with regulatory requirements or to overcome regulatory challenges. Merging with a complementary entity can sometimes provide a smoother path through regulatory approvals.

While mergers offer various strategic advantages, they also come with challenges such as integration complexities, cultural differences, and potential resistance from stakeholders. Successful mergers require careful planning, thorough due diligence, and effective post-merger integration strategies to realize the intended benefits.

An illustrative example of a prominent Indian merger is the merger between Vodafone India and Idea Cellular. Completed in 2018, this merger was a significant consolidation in the Indian telecommunications sector. Vodafone India, a subsidiary of Vodafone Group, and Idea Cellular, a part of the Aditya Birla Group, merged to create Vodafone Idea Limited, becoming one of the largest telecom operators in India. The merger aimed to enhance operational synergies, reduce costs, and strengthen the competitive position of the combined entity in the highly competitive Indian telecom market.

11.4.3 Acquisitions

Acquisitions, which involve one company purchasing another, are strategic business activities driven by a range of reasons aimed at achieving specific organizational goals. Acquisitions can take various forms, including the acquisition of assets, equity stakes, or the entire business. Here are some common reasons behind acquisitions:

1. Market Share Expansion: Acquiring another company allows a business to rapidly expand its market share. This is particularly beneficial when a company wants to strengthen its position in a specific industry or enter new markets without the time and effort required to build a market presence from scratch.

2. Product or Service Portfolio Enhancement: Companies often pursue acquisitions to enrich their product or service offerings. By acquiring a business with complementary products or services, a company can diversify its portfolio, meet evolving customer demands, and enhance its competitiveness in the marketplace.

3. Cost Synergies: Acquisitions can lead to cost synergies by eliminating redundant operations, streamlining processes, and reducing overhead expenses. Through economies of scale and operational efficiencies, the combined entity can often achieve lower per-unit costs and improved overall profitability.

4. Access to New Technologies: Acquiring companies with advanced technologies or intellectual property can provide a competitive edge. This is common in industries driven by rapid technological advancements, where acquiring innovation through acquisitions is faster than developing it internally.

5. Talent Acquisition: Acquiring a company can also mean acquiring its skilled workforce, management team, or other key personnel. This is particularly relevant when a company seeks to enhance its human capital, access specific expertise, or address talent shortages.

6. Geographical Expansion: Acquisitions offer a rapid means of expanding into new geographic regions. By acquiring companies with established operations in different locations, a business can enter new markets, navigate local regulations, and capitalize on existing customer relationships.

7. Diversification of Risk: Companies may pursue acquisitions to diversify their risk. By operating in multiple industries or markets, an organization is less vulnerable to economic downturns or disruptions in a specific sector.

8. Enhanced Financial Performance: Acquiring a financially sound company can contribute to improved financial performance. This may involve acquiring a company with strong cash flows, solid profitability, or a favorable credit rating.

9. Strategic Positioning: Acquisitions are often driven by a desire to strategically position a company in its industry. This could involve gaining a competitive advantage, responding to market trends, or preemptively addressing competitive threats.

10. Regulatory Compliance: In some cases, companies pursue acquisitions to navigate regulatory requirements more effectively. Acquiring a company with existing regulatory approvals or compliance mechanisms can simplify the regulatory process.

Successful acquisitions require careful planning, due diligence, and post-acquisition integration strategies. Understanding the strategic reasons behind an acquisition is crucial for aligning organizational goals, ensuring compatibility between the acquiring and target companies, and maximizing the value derived from the transaction.

On the international stage, a noteworthy example is the acquisition of WhatsApp by Facebook in 2014. Facebook, a leading social media giant, acquired WhatsApp, a popular messaging app, for approximately $19 billion. The acquisition was driven by Facebook's strategic objective to expand its user base and strengthen its presence in the rapidly growing mobile messaging space. WhatsApp continued to operate independently, but the acquisition provided Facebook with access to WhatsApp's extensive user base and innovative messaging platform, contributing to Facebook's overall growth and diversification strategy.

Both examples showcase the diverse motivations behind mergers and acquisitions, from market consolidation and operational synergies in the case of Vodafone Idea Limited to strategic expansion and access to new user bases and technologies in the case of Facebook's acquisition of WhatsApp. These transactions highlight the global nature of M&A activities and their role in shaping industries and markets on both regional and international scales.

11.4.4 Financial Reporting in Business Combinations
Financial reporting in business combinations involves a comprehensive and transparent presentation of the financial effects of the combination. The accounting principles for business combinations are outlined in accounting standards, such as the International

Financial Reporting Standards (IFRS) and the Generally Accepted Accounting Principles (GAAP) in the United States. The standard accounting method for business combinations is the acquisition method.

Under the acquisition method, the acquiring company recognizes the assets acquired and liabilities assumed at their fair values at the acquisition date. Goodwill is also recognized as the excess of the cost of the acquisition over the fair value of the identifiable net assets acquired. The fair value of assets and liabilities is determined through a rigorous valuation process, which includes assessments of tangible and intangible assets, liabilities, contingent liabilities, and consideration transferred.

Financial reporting for business combinations includes the preparation of consolidated financial statements. Consolidation involves combining the financial statements of the acquiring company and the acquired company as if they were a single entity. This provides a more accurate reflection of the financial position, results of operations, and cash flows of the combined entity.

The notes to the financial statements play a crucial role in explaining the details of the business combination. They typically disclose information about the purchase consideration, the fair value of assets acquired and liabilities assumed, the allocation of the purchase price, and any contingent considerations. Additionally, the notes may provide information on the expected synergies and the impact of the business combination on the financial performance and financial position of the combined entity.

Financial reporting requirements for business combinations aim to provide relevant, reliable, and comparable information to stakeholders, facilitating a better understanding of the financial implications of the transaction. Transparent and accurate reporting is essential for investors, creditors, and other stakeholders to assess the financial health, performance, and prospects of the combined entity resulting from the business combination. Compliance with accounting standards ensures consistency and comparability in financial reporting across different business combinations, contributing to the overall integrity and reliability of financial information in the corporate reporting landscape.

11.4.5 Purchase Price Allocation

Purchase Price Allocation (PPA) is a critical accounting process in the aftermath of a business combination. It involves the allocation of the total purchase price paid in an acquisition to the individual assets and liabilities acquired based on their fair values at the acquisition date. The goal of PPA is to accurately reflect the economic substance of the transaction in the financial statements and provide stakeholders with a transparent view of the acquired company's assets and liabilities.

During PPA, a thorough valuation of both tangible and intangible assets and liabilities is conducted. Tangible assets may include property, plant, equipment, and inventory, while intangible assets encompass items like trademarks, patents, customer relationships, and goodwill. Liabilities, including contingent liabilities, are also assessed. Valuation methods such as market comparisons, income approaches, and cost-based approaches are utilized to determine the fair values.

Goodwill is a significant component of PPA. It represents the excess of the purchase price over the fair value of identifiable net assets acquired. Goodwill is considered an intangible asset and is subject to impairment testing periodically.

The allocation process involves a detailed examination of the fair values of all acquired elements. The result is a detailed schedule that outlines how the total purchase price is distributed among various assets and liabilities. This schedule is then used to adjust the balance sheet of the acquiring company to reflect the fair values of the acquired assets and liabilities.

Purchase Price Allocation is not only crucial for financial reporting but also for making informed business decisions. Stakeholders, including investors, analysts, and regulators, rely on accurate and transparent financial information to assess the impact of the acquisition on the financial position and performance of the combined entity. It is essential for companies to adhere to relevant accounting standards, such as IFRS and GAAP, to ensure consistency and comparability in financial reporting across different business combinations.

11.4.6 Post-Acquisition Integration

Post-acquisition integration is a critical phase in the life cycle of a business combination, focusing on merging the operations, processes, cultures, and technologies of the acquiring and acquired entities. The success of an acquisition often hinges on how effectively and seamlessly these elements are integrated to realize the strategic objectives and synergies envisioned during the deal. The integration process typically involves a series of planned activities and initiatives aimed at achieving operational efficiency, capturing cost savings, and maximizing the overall value of the combined entity.

One crucial aspect of post-acquisition integration is cultural alignment. Companies often have distinct organizational cultures, and reconciling these differences is vital for a smooth transition. Leadership teams need to foster open communication, address concerns, and create a shared vision to build a cohesive and collaborative working environment.

Operational integration is another key focus area. This involves harmonizing business processes, consolidating redundant functions, and leveraging economies of scale to enhance efficiency. The integration team must assess and optimize the supply chain, distribution networks, and IT systems to create a unified operational structure that capitalizes on the strengths of both organizations.

Strategic alignment is essential for realizing the intended synergies. This includes aligning business strategies, combining product or service offerings, and optimizing market positioning. The integration process may also involve rationalizing product portfolios, rebranding, and creating a cohesive go-to-market strategy.

Human capital integration is a critical element, recognizing that the success of an acquisition is often contingent on the skills, talents, and engagement of the workforce. Retaining key talent, providing training and development opportunities, and ensuring a smooth transition for employees are priorities in the integration process.

Communication throughout the integration process is paramount. Clear, transparent communication helps manage uncertainties, reduce resistance, and align the entire organization with the objectives of the acquisition. This includes communicating changes in organizational structure, reporting lines, and any modifications to policies and procedures.

Post-acquisition integration is a dynamic and ongoing process that requires careful planning, execution, and monitoring. Companies need to establish dedicated integration teams, define key performance indicators, and regularly assess progress against predefined milestones. Flexibility and adaptability are crucial, as unexpected challenges may arise, requiring adjustments to the integration plan.

Ultimately, successful post-acquisition integration enhances the overall value proposition of the combined entity, creating a stronger, more competitive organization that is better positioned for sustainable growth and success in the marketplace.

11.5 Basic Principles of Management

Basic principles of management serve as fundamental guidelines for effective organizational leadership and decision-making. These principles provide a framework for managers to plan, organize, lead, and control organizational activities. Henri Fayol, a pioneering management theorist, introduced five key principles: planning, organizing, commanding, coordinating, and controlling. Planning involves setting objectives and determining the best course of action to achieve them. Organizing entails arranging resources and tasks to meet the planned objectives. Commanding involves leading and directing personnel to execute plans. Coordinating ensures that all parts of the organization work together harmoniously. Finally, controlling involves monitoring activities to ensure they align with the plans and making adjustments as needed. These principles, along with other contemporary frameworks like those by Peter Drucker and Frederick Taylor, provide a timeless foundation for effective management practices in diverse organizational settings.

11.5.1 Planning

The planning phase is a fundamental aspect of the management process, serving as the initial and crucial step in achieving organizational goals. During this phase, managers engage in a systematic and forward-thinking process of defining objectives, identifying resources, and determining the best course of action to accomplish the desired outcomes. Planning involves assessing the current state of affairs, anticipating future trends, and making decisions on how to allocate resources effectively. It requires a careful analysis of both internal and external factors that may impact the organization. The planning phase sets the direction for the entire management process, influencing subsequent actions in organizing, leading, and controlling. Successful planning involves setting clear and achievable goals, formulating strategies to achieve those goals, and developing contingency plans to address unforeseen challenges. It is a dynamic process that requires continuous evaluation and adjustment to ensure alignment with organizational priorities and changing external conditions.

11.5.2 Strategic Planning

Strategic planning is a systematic and dynamic process that organizations undertake to define their long-term objectives and develop strategies to achieve them. It involves a comprehensive analysis of internal and external factors affecting the organization, considering strengths, weaknesses, opportunities, and threats (SWOT analysis). The primary goal of strategic planning is to align an organization's resources and capabilities with the opportunities and challenges in its external environment, fostering a sustainable and competitive advantage.

The process typically begins with a clear articulation of the organization's mission, vision, and values. These elements provide a foundation for setting strategic objectives that guide decision-making throughout the organization. SWOT analysis helps identify areas where the organization can capitalize on strengths and opportunities while mitigating weaknesses and addressing potential threats.

Strategic planning often includes the development of action plans, initiatives, and projects to implement the chosen strategies. These initiatives are designed to allocate resources effectively, improve organizational performance, and position the organization for future success. The involvement of key stakeholders, including leadership, employees, and sometimes external consultants, is crucial for gathering diverse perspectives and insights during the strategic planning process.

Regular monitoring and evaluation are integral to successful strategic planning. Organizations need to assess the progress of strategic initiatives, adapt to changes in the external environment, and make adjustments to strategies as needed. Flexibility and agility are key components of effective strategic planning, allowing organizations to navigate uncertainties and capitalize on emerging opportunities.

Strategic planning is not a one-time event but rather an ongoing and iterative process that evolves with the organization and its external environment. Successful implementation of strategic plans requires strong leadership commitment, effective communication, and a collaborative organizational culture that aligns with the established strategic goals. Overall, strategic planning is a proactive and forward-looking approach that helps organizations anticipate changes, stay competitive, and achieve long-term success.

11.5.3 Operational Planning

Operational planning is a crucial component of the overall planning process within organizations, focusing on the detailed implementation of strategies and achieving short-term objectives. Unlike strategic planning, which looks at the organization as a whole and its long-term goals, operational planning zooms in on specific activities, tasks, and processes to ensure day-to-day efficiency and effectiveness. This level of planning is vital for translating the broader strategic vision into actionable steps that can be executed by different departments and teams. In operational planning, organizations break down strategic goals into manageable tasks and allocate resources accordingly. This involves creating detailed action plans, setting performance benchmarks, and establishing timelines for the completion of specific activities. Coordination among different functional areas is essential during operational planning to ensure that everyone is aligned with the organization's overall objectives. Budgeting is a significant aspect of operational planning, as it involves estimating and allocating financial resources to support the execution of operational activities. This ensures that there is a realistic financial framework in place to carry out the day-to-day functions of the organization.

Continuous monitoring and feedback mechanisms are integral to operational planning. Regular assessment of performance against predetermined benchmarks allows organizations to identify deviations, make necessary adjustments, and improve overall operational efficiency. Flexibility and adaptability are crucial in operational planning to address unforeseen challenges and changes in the business environment.

Overall, operational planning is a hands-on, practical approach that turns strategic visions into actionable steps. It ensures that the organization's resources are optimally utilized, processes are streamlined, and the workforce is aligned toward achieving short-term goals, thereby contributing to the successful implementation of the broader strategic plan.

11.5.4 Organizing

The organizing phase in management is a pivotal step where the plans developed in the earlier stages are translated into action. During this phase, managers focus on structuring resources, tasks, and activities to ensure the effective and efficient attainment of organizational goals. This involves creating an organizational structure that delineates roles, responsibilities, and reporting relationships within the workforce. Managers allocate resources such as human capital, financial assets, and technology to support the implementation of plans. Clear lines of communication and coordination are established to facilitate seamless collaboration among different departments and teams. The organizing phase seeks to create a framework that promotes efficiency, accountability, and the optimal use of available resources. By defining roles, responsibilities, and reporting structures, this phase lays the groundwork for a well-coordinated and productive work environment, enabling the organization to function smoothly and move toward the accomplishment of its objectives.

11.5.5 Organizational Structure

Organizational structure refers to the framework that defines the hierarchical arrangement of roles, responsibilities, and reporting relationships within an organization. It is a fundamental element of management and plays a crucial role in shaping how activities and tasks are organized and executed. Organizational structure is often depicted in an organizational chart, which visually represents the various departments, levels of management, and their connections.

Several common types of organizational structures exist, each with its advantages and disadvantages. A functional structure groups employees based on their specialized skills and tasks, facilitating efficiency and expertise within each functional area. Divisional structures organize employees based on products, services, or geographic locations, allowing for more focused and responsive units. Matrix structures combine aspects of both functional and divisional structures, often providing flexibility but requiring careful coordination. The choice of organizational structure depends on factors such as the organization's size, industry, culture, and strategic goals. Small, entrepreneurial firms may benefit from a flat, decentralized structure that promotes quick decision-making, while large, complex organizations may opt for a more hierarchical structure to manage diverse operations.

The organizational structure influences communication channels, decision-making processes, and the overall efficiency and effectiveness of the organization. It also impacts employee morale, as individuals' roles and responsibilities are defined by the structure. Changes in organizational strategy or external factors may necessitate adjustments to the structure to better align with new goals and challenges. Ultimately, a well-designed organizational structure enhances clarity, accountability, and the ability to adapt to changing circumstances. Regular reviews and adjustments to the structure are essential to ensure that it remains aligned with the organization's evolving needs and objectives.

11.5.6 Delegation

Delegation is a critical phase in the management process that involves entrusting authority, responsibilities, and tasks to subordinates within the organizational structure. Effective delegation is a skill that enables managers to distribute workload, empower team members, and focus on strategic priorities. During this phase, managers carefully select tasks that align with the skills, competencies, and development goals of their team members.

Delegation fosters a sense of ownership and accountability among employees, as they take on responsibilities and contribute directly to the achievement of organizational goals. It is not merely about assigning tasks but also about providing the necessary resources, guidance, and support to ensure successful task completion. Managers must communicate expectations clearly, define goals, and establish a framework for reporting and feedback.

While delegation can enhance efficiency and promote skill development, it requires a balance. Over-delegating or under-delegating can lead to issues such as employee burnout or a lack of skill development. Managers need to assess the capabilities of their team members, provide appropriate training when needed, and monitor progress to ensure tasks are completed successfully.

Delegation is closely linked to leadership and is a key factor in building a high-performing and motivated team. Effective delegation not only allows managers to focus on strategic decision-making but also promotes a culture of trust and collaboration within the organization. It empowers employees, enhances their job satisfaction, and contributes to the overall success and productivity of the team and the organization as a whole.

11.5.7 Leading

The leading phase in management is a pivotal stage where managers inspire, guide, and motivate their teams to achieve organizational goals. Leadership during this phase involves influencing and directing individuals towards the accomplishment of tasks, fostering a positive work culture, and aligning team efforts with the overall vision of the organization. Effective leaders communicate a compelling vision, set clear expectations, and provide guidance to help team members understand their roles and responsibilities. Leadership styles may vary, ranging from autocratic to democratic, depending on the organizational context and the nature of the tasks at hand. A key aspect of the leading phase is the development of strong interpersonal relationships, as leaders work to build trust, encourage collaboration, and address conflicts within the team. By creating a supportive and inspiring environment, leaders play a crucial role in enhancing employee engagement, boosting morale, and promoting a shared commitment to the organization's success.

11.5.8 Leadership Styles

- **Transformational Leadership:** Transformational leadership is a leadership style characterized by inspiring and motivating followers to achieve exceptional performance and reach their full potential. Leaders who adopt this style often create a compelling vision for the future, fostering a sense of purpose and enthusiasm among their teams. One of the key features of transformational leadership is the ability to challenge the status quo and encourage innovative thinking. Transformational leaders empower their followers by providing them with the autonomy to make decisions, fostering a culture of creativity and continuous improvement. This leadership style, popularized by leadership theorist James MacGregor Burns, emphasizes building strong relationships and trust between leaders and followers. Transformational leaders are known for their charisma, passion, and ability to communicate a compelling vision that goes beyond immediate self-interest, instilling a sense of shared values and common goals.

Transformational leadership has been associated with positive outcomes in terms of employee engagement, job satisfaction, and organizational performance. Followers under transformational leaders often feel a sense of personal growth and development as leaders invest time and resources in mentoring and coaching. Additionally, transformational leaders are adept at creating a positive organizational culture, where employees are encouraged to think creatively, take risks, and contribute meaningfully to the overall success of the organization. While transformational leadership has its strengths, it is not a one-size-fits-all approach, and its effectiveness may vary based on organizational context and individual preferences. Nonetheless, in dynamic and rapidly changing environments, transformational leadership can be a powerful force for driving innovation, adaptation, and long-term success.

- **Transactional Leadership:** Transactional leadership is a style of leadership that focuses on the exchange of rewards and punishments to motivate followers and achieve specific goals. This leadership approach is rooted in the concept of a transaction or an exchange between leaders and their subordinates. Leaders who adopt a transactional style set clear expectations, establish performance standards, and use contingent rewards or corrective actions to manage and motivate their teams.

One of the key principles of transactional leadership is the emphasis on structure and order. Leaders under this model provide explicit instructions and guidelines, defining roles, responsibilities, and performance expectations. Transactional leaders monitor adherence to these standards and intervene when deviations occur. The approach relies on the use of positive reinforcement, such as recognition, praise, or tangible rewards, to encourage desired behaviors, while corrective actions or punishments are employed in response to subpar performance. Transactional leaders often excel in stable and structured environments where routine and predictability are valued. This leadership style is effective for achieving short-term goals, maintaining operational efficiency, and ensuring compliance with established procedures. However, it may not be as well-suited for fostering innovation, creativity, or long-term organizational adaptability.

While transactional leadership has its strengths, it may be perceived as directive and controlling, potentially stifling individual initiative and intrinsic motivation. Some critics argue that a sole reliance on transactional leadership may limit the development of a positive organizational culture based on shared values and a common vision. In summary, transactional leadership provides a structured and systematic approach to managing teams, with a focus on achieving specific, well-defined objectives. While it may be effective in certain contexts, organizations that aspire to foster innovation and long-term growth often complement transactional leadership with other leadership styles that emphasize inspiration, vision, and employee development.

11.5.9 Motivation
The motivation phase in management is a crucial element in driving individual and collective efforts toward the achievement of

organizational goals. Motivation involves understanding and influencing the factors that drive people to take action, perform effectively, and contribute positively to the workplace. Managers play a pivotal role in this phase by recognizing the diverse needs and aspirations of their team members and tailoring strategies to inspire and energize them. Effective motivation goes beyond traditional methods like financial incentives and encompasses factors such as recognition, meaningful work, career development, and a positive work environment. Managers need to cultivate a motivational culture that fosters a sense of purpose, encourages collaboration, and empowers employees to take ownership of their roles. By tapping into intrinsic and extrinsic motivators, managers can create an environment where individuals are engaged, committed, and enthusiastic about their work, ultimately contributing to the overall success and productivity of the organization.

- **Intrinsic vs. Extrinsic Motivation:** In the realm of motivation, two primary forms are often discussed: intrinsic motivation and extrinsic motivation. Intrinsic motivation refers to the internal factors that drive individuals to engage in an activity or task for its own inherent rewards. This type of motivation arises from personal interest, enjoyment, or a sense of accomplishment. When individuals are intrinsically motivated, they find satisfaction in the process of the activity itself, and the activity aligns with their values and interests.

On the other hand, extrinsic motivation is driven by external factors, such as rewards, recognition, or avoiding punishment. Individuals who are extrinsically motivated are spurred into action by the desire to attain a tangible outcome or avoid a negative consequence. Extrinsic motivators can take various forms, including monetary rewards, promotions, praise, or fear of repercussions.

The interplay between intrinsic and extrinsic motivation is complex, and individuals often experience a combination of both in different contexts. The key is finding the right balance that aligns with the nature of the task or goal. While intrinsic motivation is often associated with higher levels of engagement, creativity, and job satisfaction, extrinsic motivators can be effective in certain situations, particularly for tasks that may not inherently be enjoyable but are necessary.

In workplaces, successful managers understand the nuanced dynamics of motivation and tailor their approaches accordingly. Cultivating a work environment that fosters intrinsic motivation by providing opportunities for autonomy, skill development, and meaningful work can lead to a more engaged and satisfied workforce. However, judicious use of extrinsic motivators, such as performance-based incentives or recognition programs, can also be instrumental in achieving specific organizational objectives. The challenge lies in creating a balanced motivational environment that leverages both intrinsic and extrinsic factors to optimize individual and collective performance.

- **Employee Recognition:** Employee recognition is a fundamental component of effective human resource management and organizational culture. It involves acknowledging and rewarding employees for their contributions, accomplishments, and dedication to their work. Recognition can take various forms, including verbal praise, written commendations, awards, bonuses, promotions, or other tangible rewards. The primary goal of employee recognition is to reinforce positive behaviors, boost morale, and create a workplace culture that values and appreciates the efforts of its workforce.

Recognition is a powerful motivator that goes beyond financial incentives. When employees feel acknowledged and appreciated, it contributes to their job satisfaction and overall engagement. It fosters a sense of belonging, loyalty, and commitment to the organization. Recognized employees often experience increased motivation to excel in their roles, leading to improved individual and team performance.

Effective employee recognition programs are strategic and tailored to the organization's values and objectives. They should be timely, specific, and genuine, addressing both individual and team achievements. Peer-to-peer recognition can also play a significant role in creating a positive and collaborative work environment.

In addition to its impact on employee morale, recognition contributes to talent retention and attraction. Employees who feel valued are more likely to stay with the organization, reducing turnover and recruitment costs. Moreover, a positive reputation for recognizing and appreciating employees can enhance the employer brand, making the organization more attractive to potential hires.

In summary, employee recognition is a key driver of organizational success, impacting employee engagement, satisfaction, and retention. Cultivating a culture of recognition requires a proactive approach from leadership, emphasizing the importance of appreciating and celebrating the efforts of individuals and teams. When done thoughtfully and consistently, employee recognition contributes to a positive work environment and a motivated, high-performing workforce.

11.5.10 Controlling

The controlling phase in management is a critical step in the management process that involves monitoring, evaluating, and regulating organizational activities to ensure they align with established plans and objectives. This phase is designed to measure actual performance against the predetermined standards and take corrective actions when necessary. The controlling function encompasses various aspects, including performance measurement, comparison of actual results with benchmarks, identification of deviations, and implementation of corrective measures.

In the controlling phase, managers utilize various control mechanisms such as feedback systems, performance metrics, and key performance indicators (KPIs) to assess how well the organization is progressing toward its goals. This involves collecting and analyzing data on various organizational activities, ranging from financial performance to operational efficiency and employee productivity.

Effective control measures not only identify deviations from plans but also provide insights into the root causes of these discrepancies. Managers can then take corrective actions, which may involve adjusting plans, revising processes, providing additional resources, or offering further training and development.

The controlling phase is not solely about addressing problems; it also involves recognizing and reinforcing positive performance. By acknowledging achievements and successful outcomes, managers can motivate employees and foster a culture of continuous improvement.

In summary, the controlling phase is a dynamic and ongoing process that helps organizations stay on course and adapt to changing circumstances. Through effective monitoring, evaluation, and corrective actions, managers ensure that organizational activities are in line with the established plans, facilitating the achievement of goals and objectives.

11.5.11 Performance Measurement

Performance measurement is a crucial aspect of management that involves the systematic assessment and evaluation of an organization's activities, processes, and outcomes to gauge its effectiveness and efficiency. This process aims to provide quantitative and qualitative insights into how well an organization is achieving its objectives and meeting predetermined standards. Performance measurement is integral to informed decision-making, as it allows managers to identify strengths, weaknesses, opportunities, and areas for improvement.

In the realm of management, performance measurement often involves the establishment of key performance indicators (KPIs) and metrics relevant to specific organizational goals. These indicators can span various areas, including financial performance, operational efficiency, customer satisfaction, employee productivity, and more. Well-defined KPIs provide a clear framework for assessing performance and tracking progress over time.

The choice of performance metrics depends on the nature of the organization and its strategic priorities. For example, a manufacturing company may focus on metrics related to production efficiency and defect rates, while a service-oriented organization may prioritize customer satisfaction scores and response times.

Performance measurement is not a one-time event; it is an ongoing and dynamic process. Regular monitoring and review of performance metrics allow organizations to adapt to changing circumstances, identify emerging trends, and make timely adjustments to strategies and operations. Effective performance measurement contributes to transparency, accountability, and the overall improvement of organizational processes.

Additionally, the information derived from performance measurement serves as a foundation for strategic planning and decision-making. It provides managers with insights into what is working well and where improvements are needed, guiding them in allocating resources, setting priorities, and optimizing organizational performance.

In summary, performance measurement is a vital tool in the management toolkit, offering a systematic and data-driven approach to assess and enhance organizational effectiveness. When done thoughtfully and consistently, performance measurement contributes to the achievement of organizational goals and the long-term success of the enterprise.

- **Key Performance Indicators:** Key Performance Indicators (KPIs) are quantifiable metrics used in management to evaluate and measure the performance of various aspects of an organization, business process, or individual. KPIs are essential tools for aligning organizational activities with strategic goals and objectives. By identifying and tracking key indicators, managers can gain insights into the health, efficiency, and effectiveness of different functions within the organization.

The selection of KPIs depends on the specific goals and priorities of the organization. In finance, common financial KPIs include revenue growth, profit margins, and return on investment. Operational KPIs may focus on factors like production efficiency, cycle times, or customer satisfaction. Human resources might track KPIs such as employee turnover rates, training effectiveness, or employee engagement scores.

Well-defined KPIs are characterized by clarity, relevance, and measurability. They provide a clear picture of progress toward goals, allowing for informed decision-making. Regular monitoring of KPIs facilitates early identification of trends, challenges, or areas that require improvement, enabling timely interventions.

Moreover, KPIs play a crucial role in performance management and accountability. They help communicate organizational objectives to different levels of the workforce and ensure that everyone is working toward common goals. KPIs also provide a basis for performance assessments, enabling managers to evaluate the effectiveness of strategies, initiatives, and individual contributions.

As technology and data analytics capabilities continue to advance, organizations are increasingly leveraging sophisticated tools and software to collect, analyze, and visualize KPI data. This enables real-time monitoring, enhances data accuracy, and supports more agile decision-making processes.

In summary, KPIs serve as valuable instruments in the management toolkit, aiding organizations in measuring, analyzing, and optimizing performance. When aligned with strategic objectives, KPIs provide a roadmap for success, allowing organizations to adapt, improve, and achieve sustainable growth.

- **Balanced Scorecard:** The Balanced Scorecard is a strategic management tool that provides a comprehensive framework for organizations to translate their vision and strategy into tangible objectives and performance indicators. Developed by Robert S. Kaplan and David P. Norton, the Balanced Scorecard goes beyond traditional financial metrics to incorporate key non-financial perspectives, offering a more holistic view of organizational performance.

The Balanced Scorecard typically consists of four interrelated perspectives:

1. Financial Perspective: This perspective focuses on traditional financial metrics such as revenue growth, profitability, and return on investment. While financial outcomes are crucial, the Balanced Scorecard recognizes the need for a balanced approach that considers other aspects influencing long-term success.

2. Customer Perspective: Understanding and meeting customer needs is fundamental to organizational success. The Customer Perspective involves identifying and measuring key factors that contribute to customer satisfaction, loyalty, and market share. This perspective often includes metrics related to customer service, product quality, and customer retention.

3. Internal Business Processes Perspective: This perspective emphasizes the internal processes and activities critical to delivering value to customers and achieving financial objectives. By identifying and optimizing key processes, organizations can enhance efficiency, reduce costs, and improve overall performance.

4. Learning and Growth Perspective: People and organizational capabilities play a crucial role in achieving strategic objectives. This perspective focuses on employee skills, knowledge, and innovation, as well as the organization's ability to adapt and innovate. Learning and growth metrics may include employee training, talent development, and innovation metrics.

The Balanced Scorecard serves as a strategic management system that aligns organizational activities with its vision and strategy. By incorporating multiple perspectives and performance indicators, it helps organizations avoid a myopic focus on short-term financial results. The interconnected nature of the perspectives encourages a balanced and integrated approach to decision-making, ensuring that improvements in one area do not come at the expense of others.

Implementation of the Balanced Scorecard involves defining strategic objectives, selecting relevant key performance indicators for each perspective, and regularly monitoring and reviewing performance against these metrics. The Balanced Scorecard has been widely adopted across various industries as a valuable tool for strategic planning, performance measurement, and organizational improvement.

11.5.12 Feedback and Adjustment

Feedback and adjustment are integral components of effective management and organizational improvement. Feedback involves the process of providing information, assessments, or evaluations regarding performance, processes, or outcomes. It serves as a valuable mechanism for individuals and teams to understand how well they are meeting expectations, what areas need improvement, and where they have excelled. Constructive feedback fosters a culture of continuous learning and development, empowering individuals to make informed adjustments to their work or behavior.

The adjustment phase is the responsive action taken based on the feedback received. It encompasses making changes, corrections, or improvements to align with desired outcomes or organizational goals. Effective managers and leaders recognize the importance of timely and targeted adjustments, whether it involves refining strategies, modifying processes, or addressing individual performance issues. The ability to adapt and respond to feedback demonstrates organizational agility and a commitment to continuous improvement.

Together, feedback and adjustment create a dynamic and iterative process that contributes to the overall success and resilience of an organization. By fostering a culture that values feedback, encourages open communication, and embraces the need for adjustments, organizations can enhance their ability to navigate challenges, capitalize on opportunities, and drive sustained improvement over time.

- **Continuous Improvement:** Continuous improvement, often synonymous with the concept of Kaizen in Japanese management philosophy, is a systematic and ongoing approach to enhance processes, products, or services within an organization. The core idea is to consistently seek ways to make incremental, positive changes, fostering a culture of innovation, adaptability, and excellence. Continuous improvement is not a one-time initiative but a philosophy deeply ingrained in the organizational mindset, encouraging employees at all levels to contribute ideas, identify inefficiencies, and suggest improvements.

This approach involves several key principles, including:

1. Incremental Progress: Continuous improvement emphasizes making small, incremental changes over time rather than large, disruptive transformations. This allows organizations to adapt gradually, minimizing the potential for resistance or disruptions.

2. Employee Involvement: Employees are at the forefront of continuous improvement. They are encouraged to actively participate in the identification of problems, propose solutions, and contribute to the implementation of changes. This involvement not only taps into the collective intelligence of the workforce but also fosters a sense of ownership and commitment.

3. Data-Driven Decision-Making: Continuous improvement relies on data and evidence to identify areas for enhancement. Organizations collect and analyze relevant metrics to gain insights into performance, customer feedback, and process efficiency. Data-driven decision-making ensures that improvements are targeted and aligned with strategic objectives.

4. Iterative Feedback Loops: Regular feedback loops are established to assess the impact of implemented changes. This iterative process allows organizations to learn from experience, refine strategies, and build on successes while addressing any unintended consequences or shortcomings.

Continuous improvement methodologies, such as Lean and Six Sigma, provide structured frameworks for organizations to implement and sustain this philosophy. These methodologies offer tools and techniques for process mapping, waste reduction, and problem-solving.

Embracing continuous improvement brings numerous benefits, including increased efficiency, higher quality products or services, enhanced customer satisfaction, and a more adaptable and resilient organizational culture. In a rapidly changing business environment, organizations that prioritize continuous improvement are better positioned to stay competitive, respond to market demands, and drive long-term success.

- **Adaptability:** Adaptability is a critical quality in both individuals and organizations, representing the ability to adjust, evolve, and thrive in the face of changing circumstances. In the dynamic and unpredictable landscape of today's world, adaptability is a key determinant of success. For individuals, adaptability involves a willingness to learn new skills, embrace change, and navigate uncertainty with resilience. It requires a growth mindset that sees challenges as opportunities for development.

In the organizational context, adaptability is a core component of organizational agility. Agile organizations are those that can quickly and effectively respond to shifts in the market, technological advancements, or other external factors. They foster a culture that encourages innovation, experimentation, and continuous improvement. Leaders play a crucial role in promoting adaptability by creating an environment that values flexibility, learning, and open communication.

Adaptable organizations are more likely to stay ahead of the competition, capitalize on emerging opportunities, and navigate challenges successfully. They understand that change is inevitable and view it as a constant force that can be leveraged for growth rather than resisted. This mindset allows organizations to adjust strategies, reallocate resources, and embrace new technologies or business models as needed.

In essence, adaptability is not just a response to change but a proactive stance that anticipates and prepares for it. It requires a blend of flexibility, strategic thinking, and a willingness to embrace ambiguity. Organizations and individuals that prioritize adaptability are better equipped to thrive in an ever-evolving landscape, ensuring their relevance and sustainability in the long run.

11.6 Conclusion

The intricate dance between finance and management is essential for steering organizations toward prosperity. From the nuanced world of taxation and risk management through insurance to the strategic maneuvers of business combinations and the foundational principles of effective management, each element contributes to the intricate tapestry of organizational success. As businesses navigate the ever-evolving landscape, a holistic understanding of these elements becomes imperative for leaders and decision-makers. By embracing the symbiosis of finance and management, organizations can not only weather challenges but also thrive in an era of constant change and opportunity.

Suggested Questions with Solutions

Taxation and Financial Management

Question 1: How can a company strategically plan its taxes to optimize financial performance?

Solution: A company can optimize its financial performance through tax planning by evaluating various deductions, credits, and incentives available in the tax code. Utilizing tax-efficient investment strategies, implementing cost segregation for depreciation benefits, and considering tax deferral options are essential elements. Additionally, staying informed about tax law changes and consulting with tax professionals can contribute to effective tax planning.

Insurance in Financial Management

Question 2: Discuss the role of business insurance in mitigating financial risks and ensuring stability.

Solution: Business insurance plays a vital role in financial management by providing protection against various risks such as property damage, liability claims, and business interruption. It helps ensure financial stability by transferring the financial burden of unexpected events to insurance providers. Companies should assess their specific risks and tailor insurance coverage accordingly to safeguard assets and maintain operational continuity.

Business Combinations and Financial Integration

Question 3: How can financial managers optimize value in business combinations and mergers?

Solution: Financial managers can optimize value in business combinations by conducting thorough due diligence, identifying synergies, and strategically structuring the deal. This includes assessing the financial health of the target, evaluating cost-saving opportunities, and aligning the financial and operational aspects of both entities. Additionally, effective post-merger integration is crucial for realizing synergies and maximizing financial benefits.

Basic Principles of Management and Financial Decision-Making

Question 4: Explain how the basic principles of management contribute to effective financial decision-making in organizations.

Solution: The basic principles of management, including planning, organizing, leading, and controlling, are integral to effective financial decision-making. Planning involves setting financial goals and strategies, organizing ensures optimal resource allocation, leading entails inspiring financial teams, and controlling involves monitoring and adjusting financial activities to align with organizational objectives. Applying these principles enhances the financial management process.

Synergy of Finance and International Business

Question 5: Explore the challenges and opportunities in managing international taxation for global businesses.

Solution: Managing international taxation for global businesses involves addressing challenges such as varying tax regulations, currency risks, and cross-border transactions. Opportunities lie in optimizing tax structures, leveraging tax treaties, and adopting transfer pricing strategies. Financial managers should collaborate with tax experts to navigate the complexities of international taxation and ensure compliance with relevant regulations.

Taxation Calculation

Question 6: If a company has a taxable income of $500,000 and is subject to a corporate tax rate of 25%, calculate the total amount of corporate tax payable.

Solution:

$$Corporate\ Tax\ Payable = Taxable\ Income \times Tax\ Rate$$
$$= \$500,000 \times 0.25 = \$125,000$$

Depreciation Calculation for Tax Purposes

Question 7: A business purchased machinery for $100,000. If the tax depreciation rate is 10% per year, calculate the depreciation expense for the first three years.

Solution:

$$Depreciation\ Expense$$
$$= Original\ Cost \times Depreciation\ Rate$$

$$Depreciation\ Expense = \$100,000 \times 0.10 = \$10,000$$

For the first year: $10,000
For the second year: $100,000 - $10,000 \times 2 = $80,000
For the third year: $80,000 - $10,000 = $70,000

Insurance Premium Calculation

Question 8: An individual pays an annual insurance premium of $1,200 for comprehensive coverage on their car. If the insurance company charges a monthly interest rate of 1.5% on late payments, calculate the total amount owed after being 2 months overdue.

Solution:

$$Total\ Amount\ Owed$$
$$=\ Original\ Premium$$
$$+\ Late\ Payment\ Interest$$
$$Total\ Amount\ Owed\ =\ \$1{,}200\ +\ (\$1{,}200\ \times 0.015\ \times 2)$$
$$Total\ Amount\ Owed\ =\ \$1{,}200\ +\ \$36 = \$1236$$

Tax Deduction Calculation

Question 9: If an individual's taxable income is $60,000 and they are eligible for a tax deduction of $5,000, calculate the new taxable income.

Solution:

$$New\ Taxable\ Income\ =\ Taxable\ Income\ -\ Tax\ Deduction$$
$$New\ Taxable\ Income\ =\ \$60{,}000\ -\ \$5{,}000\ =\ \$55{,}000$$

Property Tax Assessment

Question 10: The assessed value of a property is $250,000, and the property tax rate is 1.5%. Calculate the annual property tax payable.

Solution:

$$Property\ Tax\ Payable\ =\ Assessed\ Value\ \times Tax\ Rate$$
$$Property\ Tax\ Payable\ =\ \$250{,}000\ \times 0.015\ =\ \$3{,}750$$

C Code for Calculating Taxes

```c
#include <stdio.h>

// Function to calculate Income Tax
float calculateIncomeTax(float income) {
    float tax = 0.0;

    // Income tax slabs for the year 2023-24 (example values)
    if (income <= 250000) {
        tax = 0.0;
    } else if (income <= 500000) {
        tax = 0.05 * (income - 250000);
    } else if (income <= 1000000) {
        tax = 0.05 * (500000 - 250000) + 0.2 * (income - 500000);
    } else {
        tax = 0.05 * (500000 - 250000) + 0.2 * (1000000 - 500000) +
0.3 * (income - 1000000);
    }

    return tax;
}

// Function to calculate GST
float calculateGST(float amount) {
    // GST slabs (example values)
    const float GST_RATE_5 = 0.05;
    const float GST_RATE_12 = 0.12;
    const float GST_RATE_18 = 0.18;
    const float GST_RATE_28 = 0.28;

    // Assuming all amounts are inclusive of GST
    if (amount <= 1000) {
        return amount * GST_RATE_5;
    } else if (amount <= 5000) {
        return amount * GST_RATE_12;
    } else if (amount <= 10000) {
```

```
    return amount * GST_RATE_18;
  } else {
    return amount * GST_RATE_28;
  }
}

int main() {
  float income, taxableIncome, gstAmount, totalAmount;

  // Get user input for income and amount
  printf("Enter your income: $");
  scanf("%f", &income);

  printf("Enter the amount for which GST needs to be calculated:
$");
  scanf("%f", &totalAmount);

  // Calculate Income Tax
  taxableIncome = income - calculateIncomeTax(income);

  // Calculate GST
  gstAmount = calculateGST(totalAmount);

  // Display results
  printf("\nIncome Tax: $%.2f\n", calculateIncomeTax(income));
  printf("Taxable Income after deducting Income Tax: $%.2f\n",
taxableIncome);
  printf("GST Amount: $%.2f\n", gstAmount);

  return 0;
}
```

JAVA Code for Calculating Taxes

```java
import java.util.Scanner;

public class TaxCalculator {

    // Method to calculate Income Tax
    public static float calculateIncomeTax(float income) {
        float tax = 0.0f;

        // Income tax slabs for the year 2023-24 (example values)
        if (income <= 250000) {
            tax = 0.0f;
        } else if (income <= 500000) {
            tax = 0.05f * (income - 250000);
        } else if (income <= 1000000) {
            tax = 0.05f * (500000 - 250000) + 0.2f * (income - 500000);
        } else {
            tax = 0.05f * (500000 - 250000) + 0.2f * (1000000 - 500000)
+ 0.3f * (income - 1000000);
        }

        return tax;
    }

    // Method to calculate GST
    public static float calculateGST(float amount) {
        // GST slabs (example values)
        final float GST_RATE_5 = 0.05f;
        final float GST_RATE_12 = 0.12f;
        final float GST_RATE_18 = 0.18f;
        final float GST_RATE_28 = 0.28f;

        // Assuming all amounts are inclusive of GST
        if (amount <= 1000) {
            return amount * GST_RATE_5;
        } else if (amount <= 5000) {
```

```java
        return amount * GST_RATE_12;
      } else if (amount <= 10000) {
        return amount * GST_RATE_18;
      } else {
        return amount * GST_RATE_28;
      }
    }

    public static void main(String[] args) {
      Scanner scanner = new Scanner(System.in);

      System.out.print("Enter your income: $");
      float income = scanner.nextFloat();

      System.out.print("Enter the amount for which GST needs to
be calculated: $");
      float totalAmount = scanner.nextFloat();

      // Calculate Income Tax
      float taxableIncome = income - calculateIncomeTax(income);

      // Calculate GST
      float gstAmount = calculateGST(totalAmount);

      // Display results
      System.out.printf("\nIncome        Tax:        $%.2f%n",
calculateIncomeTax(income));
      System.out.printf("Taxable Income after deducting Income
Tax: $%.2f%n", taxableIncome);
      System.out.printf("GST Amount: $%.2f%n", gstAmount);

      scanner.close();
    }
  }
```

PYTHON Code for Calculating Taxes

```python
# Function to calculate Income Tax
def calculate_income_tax(income):
    # Income tax slabs for the year 2023-24 (example values)
    if income <= 250000:
        return 0.0
    elif income <= 500000:
        return 0.05 * (income - 250000)
    elif income <= 1000000:
        return 0.05 * (500000 - 250000) + 0.2 * (income - 500000)
    else:
        return 0.05 * (500000 - 250000) + 0.2 * (1000000 - 500000) + 0.3 * (income - 1000000)

# Function to calculate GST
def calculate_gst(amount):
    # GST slabs (example values)
    GST_RATE_5 = 0.05
    GST_RATE_12 = 0.12
    GST_RATE_18 = 0.18
    GST_RATE_28 = 0.28

    # Assuming all amounts are inclusive of GST
    if amount <= 1000:
        return amount * GST_RATE_5
    elif amount <= 5000:
        return amount * GST_RATE_12
    elif amount <= 10000:
        return amount * GST_RATE_18
    else:
        return amount * GST_RATE_28

# Input from the user
income = float(input("Enter your income: $"))
total_amount = float(input("Enter the amount for which GST needs to be calculated: $"))
```

```
# Calculate Income Tax
taxable_income = income - calculate_income_tax(income)

# Calculate GST
gst_amount = calculate_gst(total_amount)

# Display results
print("\nIncome Tax:
${:.2f}".format(calculate_income_tax(income)))
print("Taxable Income after deducting Income Tax:
${:.2f}".format(taxable_income))
print("GST Amount: ${:.2f}".format(gst_amount))
```

12 INDUSTRIAL RECORD-KEEPING: UNRAVELING THE ESSENCE OF FINANCIAL ACCOUNTING

12.1 Introduction

In the complex web of industrial operations, financial record-keeping stands as the bedrock of effective management. It is a meticulous process that involves capturing, classifying, and summarizing financial transactions, ensuring accuracy, transparency, and compliance. This comprehensive guide delves into the core components of industrial record-keeping, with a particular focus on the double-entry system, journal, ledger, trial balance, and cash book. By unraveling the intricacies of these fundamental aspects, organizations can enhance their financial management, make informed decisions, and navigate the challenges of the industrial landscape.

12.2 The Foundation: Double-Entry System

At the heart of industrial record-keeping lies the double-entry system, a foundational principle in accounting that traces its roots back to Luca Pacioli in the 15th century. The essence of this system is the duality of every financial transaction. In essence, for every debit, there must be an equal and corresponding credit. This dual impact ensures that the accounting equation, Assets = Liabilities + Equity, remains in balance. The double-entry system provides a robust framework for capturing the complete financial impact of transactions and serves as the cornerstone for accurate financial reporting.

12.3 Journal: The Chronological Chronicle

The journal, often considered the "book of original entry," is the first stop in the accounting journey. It is the chronological record where all financial transactions are initially entered. Each entry in the journal includes the date of the transaction, the accounts affected, a brief description, and the amounts debited and credited. This chronological organization facilitates the tracking of financial events over time and lays the groundwork for further classification and analysis.

Example Journal Entry:

Date	Account Debit	Account Credit	Description	Amount
2023-01-01	Inventory	Cash	Purchase of Raw Materials	$10,000

12.4 Ledger: The Systematic Repository

From the raw data recorded in the journal, information is classified and transferred to the ledger. The ledger is a systematic and classified record that consists of all accounts used by the organization. It is organized in a T-account format, with debits on the left and credits on the right. Each account in the ledger represents a specific asset, liability, equity, revenue, or expense. The ledger provides a detailed and organized view of an entity's financial position, laying the groundwork for the preparation of financial statements.

Example Ledger Account (Cash):

Cash Account

Date	Description	Debit ($)	Credit ($)	Balance ($)
2023-01-01	Opening Balance	-	-	10,000
2023-01-10	Sales Revenue	5,000	-	15,000
2023-01-15	Purchase of Supplies	-	2,000	13,000
2023-01-31	Closing Balance	-	-	13,000

In a Cash Account, each transaction is recorded chronologically, detailing whether it's a debit or credit, the amount involved, and the resulting balance. Here's a breakdown of the transactions:

Opening Balance (2023-01-01):

No change in the balance, as it's an opening entry.

Sales Revenue (2023-01-10):

Debit entry of $5,000 represents an increase in cash due to sales.
The new balance is $10,000 (opening balance) + $5,000 (sales)
$$= \$15,000.$$

Purchase of Supplies (2023-01-15):

Credit entry of $2,000 represents a decrease in cash due to the purchase.
The new balance is $15,000 (previous balance) - $2,000 (purchase)
$$= \$13,000.$$

Closing Balance (2023-01-31):

No change in the balance, as it's a closing entry.

This Cash Account provides a clear snapshot of the cash transactions over the specified period, allowing for easy tracking of cash inflows and outflows and providing essential information for financial management and reporting.

12.5 Trial Balance: The Preliminary Check

The trial balance is a crucial step in the accounting cycle. It is a listing of all ledger accounts and their respective debit and credit balances. The purpose of the trial balance is to ensure that the double-entry system is maintained — that the total debits equal the total credits. A balanced trial balance indicates that the ledger is arithmetically accurate, providing confidence in the accuracy of the recorded transactions.

Example Trial Balance:

Trial Balance as of 2023-01-31

Account	Debit ($)	Credit ($)	
Cash	13,000	-	
Accounts Payable	-	2,000	
Inventory	10,000	-	
Sales Revenue	-	5,000	
Supplies	-	2,000	
Total Debits	23,000	Total Credits	9,000

12.6 Cash Book: Monitoring Liquidity

The cash book is a specialized journal designed to record all cash transactions, providing a real-time snapshot of an entity's cash position. It functions as both a journal and a ledger for cash, allowing for the systematic recording of cash inflows and outflows. The cash book is particularly critical for monitoring liquidity, managing cash flows, and ensuring that the organization has the necessary funds to meet its obligations.

Example Cash Book:

Date	Description	Cash Inflow ($)	Cash Outflow ($)	Balance ($)
2023-01-01	Opening Balance	-	-	10,000
2023-01-10	Sales Revenue	5,000	-	15,000
2023-01-15	Purchase of Supplies	-	2,000	13,000
2023-01-31	Closing Balance	-	-	13,000

12.7 Conclusion: Navigating the Financial Landscape

In conclusion, industrial record-keeping is a meticulous and structured process that ensures the financial health of an organization. The double-entry system, journal, ledger, trial balance, and cash book are integral components of this process, each playing a unique role in capturing, classifying, and summarizing financial transactions. By understanding and implementing these fundamental principles, industrial entities can enhance their financial management capabilities, make informed decisions, and navigate the complexities of the industrial landscape with confidence and precision.

Suggested Questions with Solutions

Double-Entry System

Question 1: Explain the concept of the double-entry system in financial accounting and provide an example.

Solution: The double-entry system is a fundamental principle in accounting where every financial transaction has equal and opposite effects on at least two accounts. For example, when a company makes a sale, it records the revenue in the income account (credit) and the increase in cash or accounts receivable in the asset account (debit).

Journal Entries

Question 2: Create journal entries for the following transactions:
  1. Purchased raw materials on credit for $5,000.
  2. Sold finished goods for cash, realizing $8,000 in revenue.

Solution:

  1. Purchased raw materials on credit:

   - Debit Raw Materials $5,000
   - Credit Accounts Payable $5,000

  2. Sold finished goods for cash:

   - Debit Cash $8,000
   - Credit Sales Revenue $8,000

Ledger Accounts

Question 3: Develop a ledger account for "Accounts Receivable" and record transactions involving the sale of goods on credit

($10,000) and subsequent collection in cash ($7,000).

Solution:

Accounts Receivable

Date	Description	Debit ($)	Credit ($)
2023-01-01	Sale on Credit	-	10,000
2023-01-15	Cash Collection	7,000	-

Trial Balance

Question 4: Provide a trial balance using the ledger accounts for the following:
1. **Cash: $15,000**
2. **Accounts Payable: $3,000**
3. **Sales Revenue: $20,000**
4. **Rent Expense: $2,000**

Solution:

Trial Balance as of 2023-01-31

Account	Debit ($)	Credit ($)	
Cash	15,000	-	
Accounts Payable	-	3,000	
Sales Revenue	20,000	-	
Rent Expense	2,000		
Total Debits	37,000	Total Credits	3,000

Preparation of Final Accounts

Question 5: Develop a simple Trading Account for a company with the following information:

- **Opening Stock: $8,000**
- **Purchases: $25,000**
- **Sales: $40,000**
- **Closing Stock: $10,000**

Solution:

Trading Account for the year ended 2022-12-31

Particulars	Amount ($)	Particulars	Amount ($)
Opening Stock	8,000	Sales	40,000
Purchases	25,000		
Total	33,000	Total	40,000
Gross Profit	7,000		

C Code for Financial Record-Keeping

```c
#include <stdio.h>
#include <stdlib.h>

// Define a struct to represent financial transactions
struct FinancialTransaction {
    char description[100];
    float amount;
};

// Function to add a new transaction to the file
void addTransaction(FILE *file, struct FinancialTransaction
transaction) {
    fwrite(&transaction, sizeof(struct FinancialTransaction), 1, file);
}

// Function to display all transactions from the file
void displayTransactions(FILE *file) {
    struct FinancialTransaction transaction;

    rewind(file); // Move file pointer to the beginning

    printf("\nFinancial Transactions:\n");
    while (fread(&transaction, sizeof(struct FinancialTransaction), 1,
file) == 1) {
        printf("Description:     %s,     Amount:     %.2f\n",
transaction.description, transaction.amount);
    }
}

int main() {
    FILE *transactionFile;
    struct FinancialTransaction newTransaction;

    // Open the file in binary read-write mode
    transactionFile = fopen("financial_records.dat", "ab+");
```

```c
        if (transactionFile == NULL) {
            fprintf(stderr, "Error opening file for financial
transactions.\n");
            return 1;
        }

        int choice;

        do {
            // Display menu
            printf("\n1. Add Transaction\n");
            printf("2. Display Transactions\n");
            printf("3. Exit\n");

            printf("Enter your choice: ");
            scanf("%d", &choice);

            switch (choice) {
                case 1:
                    // Add a new transaction
                    printf("Enter Transaction Description: ");
                    scanf("%s", newTransaction.description);

                    printf("Enter Transaction Amount: ");
                    scanf("%f", &newTransaction.amount);

                    addTransaction(transactionFile, newTransaction);
                    printf("Transaction added successfully.\n");
                    break;

                case 2:
                    // Display all transactions
                    displayTransactions(transactionFile);
                    break;

                case 3:
                    // Exit the program
                    printf("Exiting the program. Goodbye!\n");
```

```
            break;

        default:
            printf("Invalid choice. Please enter a valid option.\n");
    }

} while (choice != 3);

// Close the file before exiting
fclose(transactionFile);

return 0;
}
```

PYTHON Code for Financial Record-Keeping

```python
import pickle

# Define a class to represent financial transactions
class FinancialTransaction:
    def __init__(self, description, amount):
        self.description = description
        self.amount = amount

# Function to add a new transaction to the file
def add_transaction(file, transaction):
    transactions = load_transactions(file)
    transactions.append(transaction)
    save_transactions(file, transactions)

# Function to display all transactions from the file
def display_transactions(file):
    transactions = load_transactions(file)

    print("\nFinancial Transactions:")
    for transaction in transactions:
        print(f"Description: {transaction.description}, Amount: {transaction.amount:.2f}")

# Function to load transactions from the file
def load_transactions(file):
    try:
        with open(file, 'rb') as f:
            transactions = pickle.load(f)
    except (FileNotFoundError, EOFError):
        transactions = []
    return transactions

# Function to save transactions to the file
def save_transactions(file, transactions):
    with open(file, 'wb') as f:
```

```
        pickle.dump(transactions, f)

def main():
    transaction_file = "financial_records.pkl"

    while True:
        # Display menu
        print("\n1. Add Transaction")
        print("2. Display Transactions")
        print("3. Exit")

        choice = input("Enter your choice: ")

        if choice == "1":
            # Add a new transaction
            description = input("Enter Transaction Description: ")
            amount = float(input("Enter Transaction Amount: "))
            new_transaction    =    FinancialTransaction(description,
amount)
            add_transaction(transaction_file, new_transaction)
            print("Transaction added successfully.")
        elif choice == "2":
            # Display all transactions
            display_transactions(transaction_file)
        elif choice == "3":
            # Exit the program
            print("Exiting the program. Goodbye!")
            break
        else:
            print("Invalid choice. Please enter a valid option.")

if __name__ == "__main__":
    main()
```

13 PREPARATION OF FINAL ACCOUNTS: UNVEILING THE FINANCIAL CANVAS

𝕱inancial reporting is the culmination of meticulous record-keeping and analysis, providing stakeholders with a comprehensive view of an organization's performance and financial position. The preparation of final accounts, including the Trading Account, Profit and Loss Account, and Balance Sheet, is a crucial step in this process. In this extensive exploration, we delve into each component, unraveling the intricacies of their preparation and conducting a simple study of a balance sheet.

13.1 Trading Account: Unveiling the Economic Activities

A Trading Account is a financial statement that unveils the economic activities related to buying and selling goods and services within a specific period. It provides a detailed record of revenues, costs of goods sold, and gross profit, offering insights into a company's core operational activities. Analyzing a Trading Account helps stakeholders assess the efficiency and profitability of a business's trading operations.

13.1.1 Introduction to Trading Account

The Trading Account is the first financial statement in the preparation of final accounts. It serves as a mirror reflecting the economic activities related to the purchase and sale of goods. The primary goal is to determine the gross profit or loss generated by these trading activities.

13.1.2 Structure of the Trading Account

The Trading Account follows a straightforward structure, showcasing the direct costs associated with the production or acquisition of goods. It consists of two main sections: the debit side (for opening stock and purchases) and the credit side (for sales and closing stock).

Example of Trading Account:

Trading Account for the year ended 31ˢᵗ December 2022

Particulars	Amount ($)	Particulars	Amount ($)
Opening Stock	10,000	Sales	70,000
Purchases	40,000		
Total (Debit side)	50,000	Total (Credit side)	70,000
Gross Profit		20,000	

In this example, the opening stock is combined with purchases on the debit side, and sales are recorded on the credit side. The difference between the total sales and the combined total of opening stock and purchases represents the gross profit.

Example: Let's go through an example of preparing a trading account from a trial balance. For simplicity, let's assume a fictional company called ABC Trading Co.

Trial Balance of ABC Trading Co. as of December 31, 2022

Particulars	Amount ($)
Opening Stock	20,000
Cash Purchases	30.000
Credit Purchase	50,000
Cash Sales	1,50,000
Credit Sales	2,50,000
Closing Stock	60,000

Trading Account of ABC Trading Co. as of December 31, 2022

Dr.	$		$	Cr.
To Opening Stock	20,000	By Cash Sales 1,50,000		
To Cash Purchases 30,000		By Credit Sales 2,50,000		
To Credit Purchases 50,000		By Total Sales	4,00,000	
To Total Purchases	80,000	By Closing Stock	60,000	
To Gross Profit c/d	3,60,000			
	4,60,000		4,60,000	

By Gross Profit b/d $12,000

This gross profit will be transferred to the Profit and Loss Account, and further adjustments for other expenses will be made to determine the net profit for the period.

13.2 Profit and Loss Account: Analyzing Operational Performance

The Profit and Loss Account, often referred to as the income statement, is a crucial financial statement that provides a comprehensive overview of a company's operational performance over a specific period. This statement is designed to showcase the revenue generated and the expenses incurred during that time frame, ultimately revealing whether the company has made a profit or incurred a loss. By analyzing the Profit and Loss Account, stakeholders gain insights into the efficiency of a business's core operations. It highlights the company's ability to generate revenue, manage costs, and ultimately achieve profitability. The key components of the statement include revenue from sales, cost of goods sold (COGS), gross profit, operating expenses, and net profit or loss. Stakeholders, including investors, creditors, and management, utilize this financial tool to assess the company's financial health and make informed decisions about its future prospects and strategies. A positive net profit indicates successful operational performance, while a net loss may necessitate a closer examination of cost management and revenue generation strategies.

13.2.1 Introduction to Profit and Loss Account

Following the Trading Account, the Profit and Loss Account takes center stage in evaluating the operational performance of a business. It encompasses all revenues and expenses, ultimately revealing the net profit or loss.

13.2.2 Structure of the Profit and Loss Account

The Profit and Loss Account, also known as the Income Statement, is a crucial financial statement that provides a snapshot of a company's revenues, expenses, gains, and losses over a specific period, typically a fiscal quarter or year. The structure of the Profit and Loss Account follows a systematic format to present a clear and concise overview of a company's financial performance. It begins with the revenue section, where sales, fees, and other income sources are listed, representing the top line of the statement. Direct costs or the cost of goods sold (COGS) are then deducted from revenues to arrive at the gross profit. Following this, operating expenses, such as selling, general and administrative expenses, are subtracted to obtain the operating profit.

Further down the statement, non-operating items, gains, and losses are included, leading to the calculation of the net profit before tax. Finally, income tax expenses are deducted to arrive at the net profit, reflecting the bottom line of the Profit and Loss Account. This structured presentation allows stakeholders to assess the company's profitability and financial health comprehensively.

Example of Profit and Loss Account:

Mike Cycles			
Profit and Loss Statement			
For the year ended 31ˢᵗ December 2022			
Income			
	Sales	$4,00,000	(10,000 cycles @ $ 40 each)
	Total Sales	$4,00,000	
Cost of Goods Sold			
	Opening Stock	Nil	
	Stock Purchases	$3,30,000	
	Less Closing Stock	$30,000	
Total Cost of Goods Sold (COGS)		$3,00,000	(See note)
Gross Profit		$1,00,000	
Expenses			
	Advertising	$6,000	
	Bank Service Charges	$2,000	
	Insurance	$4,000	
	Payroll	$28,000	
	Professional Fees (Legal, Accounting)	$4,000	
	Utilities & Telephone	$2,000	
	Other: Computer Software	$4,000	
	Expenses total	$50,000	
Net Profit before Tax		**$50,000**	

Note; Cost of Goods Sold calculation:		
Towards the end of the year, Mike Cycles manages to purchase 1000 more cycles on credit from his supplier for an order in the new year. This leaves them with $30,000 of stock on hand at the end of the year.		
Mike Cycles' Cost of Goods Calculation		
Opening Stock		Nil
Add Stock Purchased during the year	$3,30,000	(11000 cycles @ $30 each)
Equals Stock available to sell	$3,30,000	
Less Stock on hand at end of year	$30,000	(1000 cycles @ $30 each)
Cost of Goods Sold	**$3,00,000**	

13.3 Balance Sheet: Portraying Financial Position

The Balance Sheet, also known as the Statement of Financial Position, is a fundamental financial statement that provides a snapshot of a company's financial position at a specific point in time. It is structured in a way that reflects the accounting equation: Assets = Liabilities + Equity. The Balance Sheet is divided into two main sections. On the left side, assets are listed in order of liquidity, with current assets like cash, accounts receivable, and inventory presented first, followed by non-current assets such as property, plant, and equipment. On the right side, liabilities and equity are detailed. Current liabilities, representing short-term obligations, come first, followed by long-term liabilities and shareholders' equity. Liabilities include items like accounts payable and long-term debt, while equity encompasses common stock, retained earnings, and additional paid-in capital. The Balance Sheet offers a comprehensive view of a company's financial health by illustrating its resources (assets) and how those resources are financed (liabilities and equity), aiding investors, creditors, and management in assessing the firm's stability and solvency.

13.3.1 Introduction to Balance Sheet

The Balance Sheet is the final chapter in the trilogy of final accounts. It provides a snapshot of an organization's financial position at a specific point in time, presenting its assets, liabilities, and equity.

13.3.2 Structure of the Balance Sheet

The Balance Sheet is divided into two main sections: the assets side (listing assets in order of liquidity) and the liabilities side (depicting both short-term and long-term liabilities along with the owner's equity).

Example of Balance Sheet:

XYZ InfoTech Company			
Balance Sheet			
As of 05/01/2023			
All Amounts are in USD			
	Current Period	Prior Period	Increase (Decrease)
	05/01/22 to 05/01/23	05/01/21 to 05/01/22	05/01/22 to 05/01/23
ASSETS			
Current Assets:			
Cash	40,500.00	10,000.00	30,500.00
Petty Cash	300.00	100.00	200.00
Accounts Receivables	5,010.00	4,000.00	1,010.00
Inventory	20,880.00	21,000.00	(120.00)
Prepaid Expenses	1,200.00	1,100.00	100.00
Employee Advances	200.00	-	200.00
Temporary Investments	-	-	-
Total Current Assets	68,090.00	36,200.00	31,890.00
Fixed Assets:			
Land	2,50,000.00	1,00,000.00	1,50,000.00
Buildings	60,200.00	59,235.00	965.00
Furniture and Equipment	28,000.00	30,000.00	(2,000.00)
Computer Equipment	5,100.00	5,200.00	(100.00)
Vehicles	32,500.00	31,000.00	1,500.00
Less: Accumulated Depreciation	(43,183.00)	(45,201.00)	2,018.00
Total Fixed Assets	3,32,617.00	1,80,234.00	1,52,383.00

Other Assets:						
Trademarks	4,900.00		4,100.00		800.00	
Patents	800.00		500.00		300.00	
Security Deposits	3,500.00		1,900.00		1,600.00	
Other Assets	583.00		600.00		(17.00)	
Total Other Assets	9,783.00		7,100.00		2,683.00	
					-	
TOTAL ASSETS	**4,10,490.00**		**2,23,534.00**		**1,86,956.00**	
LIABILITIES						
Current Liabilities:						
Accounts Payable	18,200.00		18,000.00		200.00	
Business Credit Cards	7,500.00		6,000.00		1,500.00	
Sales Tax Payable	1,080.00		1,000.00		80.00	
Payroll Liabilities	2,600.00		2,500.00		100.00	
Other Liabilities	900.00		800.00		100.00	
Current Portion of Long-Term Debt	12,100.00		12,000.00		100.00	
Total Current Liabilities	42,380.00		40,300.00		2,080.00	
Long-Term Liabilities:						
Notes Payable	50,100.00		48,500.00		1,600.00	
Mortgage Payable	1,48,000.00		1,46,231.00		1,769.00	
Less: Current portion of Long-term debt	(1,000.00)		(1,200.00)		200.00	
Total Long-Term Liabilities	1,97,100.00		1,93,531.00		3,569.00	
EQUITY						
Capital Stock/Partner's Equity	21,000.00		20,000.00		1,000.00	
Opening Retained Earnings	25,530.00		23,000.00		2,530.00	
Dividends Paid/Owner's Draw	(5,000.00)		(5,000.00)		-	
Net Income (Loss)	12,000.00		7,000.00		5,000.00	
Total Equity	53,530.00		45,000.00		8,530.00	
TOTAL LIABILITIES & EQUITY	**2,93,010.00**		**2,78,831.00**		**14,179.00**	

In this representation, the assets side includes current assets like cash and bank, and accounts receivable, along with fixed assets like property and equipment. The liabilities side encompasses both current liabilities, such as accounts payable and short-term loans, and long-term liabilities, such as long-term loans. The owner's equity represents the residual interest in the assets after deducting liabilities.

13.4. A Simple Study of Balance: Interpreting the Financial Health

A simple study of a company's balance sheet is instrumental in interpreting its financial health and stability. By examining the balance between assets, liabilities, and equity, stakeholders can gain valuable insights into the firm's overall financial position. A healthy balance sheet typically exhibits a stable relationship between assets and liabilities, ensuring that the company can meet its short-term and long-term obligations. If a company carries a higher proportion of current assets relative to current liabilities, it signifies liquidity and the ability to cover immediate financial commitments. On the other hand, an excessive reliance on debt can raise concerns about financial leverage and long-term sustainability. Moreover, the equity section reflects the owners' stake in the business and its retained earnings, showcasing the company's profitability and capacity for reinvestment. Overall, a careful analysis of a balance sheet allows investors, creditors, and management to gauge the financial soundness and resilience of a company, aiding in informed decision-making and risk assessment.

13.4.1 Overview of the Balance Sheet

The Balance Sheet is not merely a compilation of numbers; it is a dynamic document that provides insights into a company's financial health. A simple study involves analyzing key components to gauge liquidity, solvency, and overall financial stability.

13.4.2 Interpreting Key Elements

Interpreting key elements in financial statements is a critical skill for stakeholders seeking a comprehensive understanding of a company's performance and financial health. Each financial statement, including the Income Statement, Balance Sheet, and Cash Flow Statement, contains key elements that convey valuable information. In the Income Statement, revenue, expenses, and net profit provide insights into the company's operational efficiency and profitability. The Balance Sheet's assets, liabilities, and equity showcase the firm's financial position and capital structure, revealing its ability to meet obligations. Analyzing the Cash Flow Statement helps assess the company's ability to generate cash and manage liquidity. Ratios derived from these key elements, such as profitability ratios, liquidity ratios, and leverage ratios, further assist in evaluating performance and making informed decisions. Successful interpretation of these elements requires a holistic view, considering industry benchmarks, historical trends, and the company's overall strategic context. Ultimately, a nuanced understanding of these key elements empowers stakeholders to make sound financial decisions and navigate the complexities of the business landscape.

Current Assets and Liabilities:
- **Current Assets (Cash, Bank, Accounts Receivable):** Indicate the company's liquidity and ability to meet short-term obligations.
- **Current Liabilities (Accounts Payable, Short-term Loans):** Reflect the short-term obligations that the company needs to settle.

Fixed Assets:
- **Property and Equipment:** Represent long-term investments, contributing to the company's operational capacity.

Long-term Liabilities:
- **Long-term Loans:** Reflect the company's ability to manage long-term debt, contributing to its solvency.

Owner's Equity:
- **Owner's Equity:** Represents the residual interest of the owner in the company's assets after settling liabilities.

13.4.3 Financial Ratios
Financial ratios are powerful tools that enable stakeholders to assess a company's performance, profitability, liquidity, and overall financial health. These ratios are derived from the financial statements and provide valuable insights when analyzed in conjunction with each other. Some key financial ratios, along with their mathematical formulae, are instrumental in financial analysis.

Profitability Ratios:

- **Return on Equity (ROE):** This ratio measures the company's ability to generate profit from shareholders' equity. The formula is

ROE = Net Income / Shareholders' Equity

- **Gross Profit Margin:** Calculated as (Gross Profit / Revenue), this ratio indicates the percentage of revenue retained after accounting for the cost of goods sold.

Liquidity Ratios:

- **Current Ratio:** This ratio assesses a company's ability to cover short-term liabilities with its short-term assets. The formula is Current

Ratio = Current Assets / Current Liabilities

- **Quick Ratio (Acid-Test Ratio):** Similar to the current ratio but more stringent, it excludes inventory from current assets.

Quick Ratio = (Current Assets - Inventory) / Current Liabilities

Efficiency Ratios:

- **Inventory Turnover:** This ratio measures how efficiently a company manages its inventory. The formula is

Inventory Turnover = Cost of Goods Sold / Average Inventory

- **Accounts Receivable Turnover:** Indicates how quickly a company collects cash from customers. The formula is Accounts

Receivable Turnover = Net Sales / Average Accounts Receivable

Solvency Ratios:

- **Debt-to-Equity Ratio:** This ratio assesses the proportion of debt used to finance the company's assets relative to shareholders' equity.

Debt-to-Equity Ratio = Total Debt / Shareholders' Equity

- **Interest Coverage Ratio:** It gauges a company's ability to cover interest expenses with its earnings before interest and taxes. The formula is

Interest Coverage Ratio = Earnings Before Interest and Taxes (EBIT) / Interest Expense

These are just a few examples, and there are numerous other ratios that cater to specific aspects of a company's financial performance. Interpreting these ratios involves comparing them to industry benchmarks, historical data, and analyzing trends over time, providing a comprehensive picture of a company's financial situation. Utilizing mathematical formulae for these ratios enhances precision and objectivity in financial analysis.

13.5 Conclusion: The Culmination of Financial Insight

In conclusion, the preparation of final accounts is not just a routine financial exercise; it is a journey that unveils the economic narrative of a business. The Trading Account illuminates the dynamics of buying and selling goods, the Profit and Loss Account dissects operational performance, and the Balance Sheet portrays the financial snapshot of an organization. Through a simple study of the balance sheet, stakeholders gain valuable insights into liquidity, solvency, and overall financial stability. By comprehending and interpreting these final accounts, organizations can not only meet regulatory requirements but also make informed decisions, foster transparency, and navigate the complex landscape of financial management with confidence.

Suggested Questions with Solutions

Question 1:
Scenario: ABC Corporation is preparing its final accounts for the year. Explain the importance of disclosing contingent liabilities in the notes to the financial statements and provide an example.

Solution: Contingent liabilities are potential obligations that may arise in the future, depending on the outcome of uncertain events. It is crucial to disclose them in the notes to the financial statements to provide transparency and help stakeholders assess the company's risk exposure. For instance, ABC Corporation might have provided guarantees for loans on behalf of its subsidiaries. Disclosing such contingent liabilities ensures that users of financial statements are aware of the potential financial impact.

Question 2:
Scenario: XYZ Ltd. is a manufacturing company. Discuss the ethical considerations involved in choosing the depreciation method for its assets and how it can impact the company's financial statements.

Solution: Choosing a depreciation method involves ethical considerations as it affects the allocation of expenses over time. The company must consider factors such as the expected useful life of assets and the method's impact on reported profits. For example, using a more accelerated method might result in higher depreciation expenses initially, reducing reported profits and tax liabilities. Ethical considerations involve presenting a true and fair view of financial performance and avoiding manipulation for short-term gains.

Question 3:
Scenario: LMN Corporation sold a significant portion of its assets during the year. Explain how the gain or loss on the sale of assets is reflected in the final accounts and its implications.

Solution: The gain or loss on the sale of assets is reflected in the Trading and Profit and Loss Account. A gain increases the net profit, while a loss reduces it. This can impact the overall financial health and profitability of the company. For instance, if LMN Corporation sold assets at a gain, it would positively contribute to its net profit, potentially improving its financial position. Conversely, a loss might indicate challenges or the need for strategic adjustments.

Question 4:
Scenario: PQR Ltd. has outstanding legal disputes that may lead to financial obligations. Discuss how the uncertainty surrounding legal contingencies is handled in the preparation of final accounts.

Solution: Legal contingencies represent potential liabilities arising from pending lawsuits or disputes. In the preparation of final accounts, companies disclose information about such contingencies in the notes to the financial statements. The disclosure includes a description of the nature of the legal issues, potential financial impact, and the company's assessment of the likelihood of an unfavorable outcome. This ensures transparency and helps stakeholders understand the company's exposure to legal risks.

Question 5:
Scenario: UVW Ltd. is a technology company with a significant investment in research and development. Discuss the accounting treatment of research and development costs and their impact on the company's financial statements.

Solution: Research and development costs are critical for technology companies, impacting innovation and future growth. In financial statements, these costs are typically expensed as incurred. However, certain development costs, if meeting specific criteria, may be capitalized. Capitalization involves recognizing these costs as assets on the balance sheet, amortizing them over their useful life. The choice between expensing and capitalizing R&D costs can significantly affect reported profits and asset values, influencing investors' perceptions of the company's financial health and potential for future earnings.

Question 6:
Scenario: LMN Corporation provided the following information for the year ended December 31, 2023:

- **Opening Stock: $12,000**
- **Purchases: $50,000**
- **Sales: $80,000**
- **Closing Stock: $18,000**
- **Operating Expenses: $15,000**

Calculate the Gross Profit and Net Profit for the year.

Solution: The Gross Profit is calculated as the difference between Sales and Cost of Goods Sold (COGS). COGS is determined by subtracting Closing Stock from the sum of Opening Stock and Purchases.

COGS = Opening Stock + Purchases - Closing Stock
Gross Profit = Sales - COGS
The Net Profit is then obtained by subtracting Operating Expenses from Gross Profit.

Net Profit = Gross Profit - Operating Expenses
COGS = $12,000 + $50,000 - $18,000 = $44,000
Gross Profit = $80,000 - $44,000 = $36,000
Net Profit = $36,000 - $15,000 = $21,000

Question 7:
Scenario: ABC Ltd. has the following balances on December 31, 2023:

- **Capital: $120,000**
- **Drawings: $10,000**
- **Purchases: $60,000**
- **Sales: $90,000**

- **Salaries: $18,000**
- **Rent: $3,000**
- **Carriage Inwards: $1,500**
- **Carriage Outwards: $1,000**
- **Discount Allowed: $2,000**
- **Discount Received: $1,200**
- **Returns Inward: $5,000**
- **Returns Outward: $3,000**
- **Bad Debts: $1,500**
- **Opening Stock: $15,000**
- **Debtors: $20,000**
- **Creditors: $8,000**
- **Cash: $4,000**
- **Bank: $6,000**

Calculate the Net Profit and determine the closing capital for the year.

Solution: To calculate the Net Profit and determine the closing capital for the year, let's follow these steps:

1. Calculate the Cost of Goods Sold (COGS) using the formula:
COGS = Opening Stock + Purchases - Returns Inward

2. Calculate the Gross Profit using the formula:
Gross Profit = Sales - COGS

3. Calculate the Net Profit using the formula:
Net Profit = Gross Profit - (Salaries + Rent + Carriage Outwards + Discount Allowed + Returns Outward)

4. Determine the closing capital using the formula:
Closing Capital = Opening Capital + Net Profit - Drawings

Now, let's perform the calculations:

COGS = $15,000 + $60,000 - $5,000 = $70,000

Gross Profit = $90,000 - $70,000 = $20,000

Net Profit = $20,000 - ($18,000 + $3,000 + $1,000 + $2,000 + $3,000)
    = $20,000 - $27,000
    = -$7,000 (Note: A negative net profit indicates a loss)

Closing Capital = $120,000 - $7,000 - $10,000 = $103,000

Therefore, the Net Profit for the year is -$7,000 (indicating a loss), and the closing capital is $103,000. The final value of Closing Capital would depend on the calculated value of COGS.

C Code for Preparation of Final Accounts

```c
#include <stdio.h>

int main() {
    // Declarations
    float openingCapital, drawings, purchases, sales, salaries, rent,
carriageInwards, carriageOutwards;
    float discountAllowed, discountReceived, returnsInward,
returnsOutward, badDebts, openingStock;
    float debtors, creditors, cash, bank;

    // Input financial information
    printf("Enter Opening Capital: $");
    scanf("%f", &openingCapital);

    printf("Enter Drawings: $");
    scanf("%f", &drawings);

    printf("Enter Purchases: $");
    scanf("%f", &purchases);

    printf("Enter Sales: $");
    scanf("%f", &sales);

    // ... (similarly, input other financial information)

    // Calculate Cost of Goods Sold (COGS)
    float cogs = openingStock + purchases - returnsInward;

    // Calculate Gross Profit
    float grossProfit = sales - cogs;

    // Calculate Net Profit
    float operatingExpenses = salaries + rent + carriageOutwards +
discountAllowed + returnsOutward;
    float netProfit = grossProfit - operatingExpenses;
```

```
// Determine Closing Capital
float closingCapital = openingCapital + netProfit - drawings;

// Display Results
printf("\n=== FINAL ACCOUNTS ===\n");
printf("Net Profit: $%.2f\n", netProfit);
printf("Closing Capital: $%.2f\n", closingCapital);

return 0;
}
```

JAVA Code for Preparation of Final Accounts

```java
import java.util.Scanner;

public class FinalAccounts {

    public static void main(String[] args) {
        Scanner scanner = new Scanner(System.in);

        // Declarations
        float openingCapital, drawings, purchases, sales, salaries, rent,
carriageInwards, carriageOutwards;
        float discountAllowed, discountReceived, returnsInward,
returnsOutward, badDebts, openingStock;
        float debtors, creditors, cash, bank;

        // Input financial information
        System.out.print("Enter Opening Capital: $");
        openingCapital = scanner.nextFloat();

        System.out.print("Enter Drawings: $");
        drawings = scanner.nextFloat();

        System.out.print("Enter Purchases: $");
        purchases = scanner.nextFloat();

        System.out.print("Enter Sales: $");
        sales = scanner.nextFloat();

        // ... (similarly, input other financial information)

        // Calculate Cost of Goods Sold (COGS)
        float cogs = openingStock + purchases - returnsInward;

        // Calculate Gross Profit
        float grossProfit = sales - cogs;

        // Calculate Net Profit
```

```java
        float operatingExpenses = salaries + rent + carriageOutwards
+ discountAllowed + returnsOutward;
        float netProfit = grossProfit - operatingExpenses;

        // Determine Closing Capital
        float closingCapital = openingCapital + netProfit - drawings;

        // Display Results
        System.out.println("\n=== FINAL ACCOUNTS ===");
        System.out.printf("Net Profit: $%.2f%n", netProfit);
        System.out.printf("Closing        Capital:        $%.2f%n",
closingCapital);

        scanner.close();
    }
}
```

PYTHON Code for Preparation of Final Accounts

```python
# Input financial information
opening_capital = float(input("Enter Opening Capital: $"))
drawings = float(input("Enter Drawings: $"))
purchases = float(input("Enter Purchases: $"))
sales = float(input("Enter Sales: $"))
salaries = float(input("Enter Salaries: $"))
rent = float(input("Enter Rent: $"))
carriage_inwards = float(input("Enter Carriage Inwards: $"))
carriage_outwards = float(input("Enter Carriage Outwards: $"))
discount_allowed = float(input("Enter Discount Allowed: $"))
discount_received = float(input("Enter Discount Received: $"))
returns_inward = float(input("Enter Returns Inward: $"))
returns_outward = float(input("Enter Returns Outward: $"))
bad_debts = float(input("Enter Bad Debts: $"))
opening_stock = float(input("Enter Opening Stock: $"))
debtors = float(input("Enter Debtors: $"))
creditors = float(input("Enter Creditors: $"))
cash = float(input("Enter Cash: $"))
bank = float(input("Enter Bank: $"))

# Calculate Cost of Goods Sold (COGS)
cogs = opening_stock + purchases - returns_inward

# Calculate Gross Profit
gross_profit = sales - cogs

# Calculate Net Profit
operating_expenses = salaries + rent + carriage_outwards + discount_allowed + returns_outward
net_profit = gross_profit - operating_expenses

# Determine Closing Capital
closing_capital = opening_capital + net_profit - drawings

# Display Results
print("\n=== FINAL ACCOUNTS ===")
print(f"Net Profit: ${net_profit:.2f}")
print(f"Closing Capital: ${closing_capital:.2f}")
```

14 INDUSTRIAL COSTS: A COMPREHENSIVE ANALYSIS

14.1 Introduction

Industrial costs are at the core of a company's financial landscape, influencing its profitability, operational efficiency, and long-term sustainability. In this comprehensive exploration, we will delve into the classification of industrial costs, focusing on material cost control, labour cost control, overhead cost control, and the quantitative aspects of depreciation and replacement studies.

14.2 Classification of Industrial Costs

The classification of industrial costs is a vital aspect of financial management that provides a structured framework for understanding and analyzing the various expenses incurred in the production and operation of goods and services. Industrial costs are typically categorized into three primary classes: material costs, labour costs, and overhead costs. Material costs encompass the expenses related to acquiring raw materials, components, and supplies essential for the production process. Labour costs include wages, salaries, benefits, and other expenses associated with the workforce. Overhead costs comprise all indirect expenditures necessary for overall business operations, such as utilities, rent, maintenance, and administrative expenses. This classification system allows businesses to allocate resources effectively, implement targeted cost control measures, and gain insights into the specific components contributing to the overall cost structure. It serves as a foundation for more in-depth financial analysis, enabling organizations to optimize their cost management strategies and enhance operational efficiency.

14.2.1 Material Cost Control

Material costs represent the expenses incurred in acquiring raw materials, components, and supplies essential for the production process. Material cost control involves strategic procurement and efficient inventory management. The mathematical formula for calculating material costs is straightforward:

Material Cost = Opening Stock + Purchases - Closing Stock

Efficient material cost control requires minimizing waste and optimizing inventory turnover. Just-in-time (JIT) inventory systems are often employed, emphasizing a precise balance between procurement and consumption.

14.2.2 Labour Cost Control

Labour costs encompass wages, salaries, benefits, and other expenses related to the workforce. Effective labour cost control involves optimizing staffing levels and improving workforce productivity. The mathematical formula for calculating labour costs is:

$$Labour\ Cost = (Number\ of\ Hours\ Worked \times Hourly\ Wage) + Overtime\ Pay + Benefits$$

Optimizing labour costs requires strategies such as performance-based incentives and technology adoption to enhance efficiency. The goal is to maximize output while minimizing labour-related expenses.

14.2.3 Overhead Cost Control

Overhead costs include indirect expenses necessary for business operations. Effective overhead cost control involves streamlining administrative and operational processes. The mathematical formula for overhead cost calculation is:

Overhead Cost = Fixed Overhead + Variable Overhead

Where:
Fixed Overhead = Rent + Insurance + Depreciation +.....
Variable Overhead = Utilities + Maintenance + Supplies + ...

Controlling overhead costs requires meticulous budgeting, variance analysis, and the elimination of non-essential expenses.

14.3 Depreciation and Replacement Studies

Depreciation and replacement studies are critical components of strategic asset management within an organization. Depreciation is the systematic allocation of the cost of tangible assets over their useful life. This process involves spreading the initial cost of an asset over time, reflecting its diminishing value due to factors such as wear and tear, obsolescence, or technological advancements. The goal is to match the cost of the asset with the revenue it generates during its operational life. Replacement studies, on the other hand, focus on evaluating when and how assets should be replaced. These studies consider factors such as the current state of assets, technological advancements, changing business requirements, and economic considerations. The Total Cost of Ownership (TCO) is a key metric in replacement studies, encompassing not only the initial purchase cost but also operating, maintenance, and disposal costs throughout the asset's lifecycle. Together, depreciation and replacement studies provide organizations with valuable insights for making informed decisions about managing their assets efficiently, optimizing costs, and ensuring long-term sustainability.

14.3.1 Depreciation

Depreciation is the systematic allocation of the cost of tangible assets over their useful life. Several methods are commonly used to calculate depreciation, each offering a different approach to allocating the cost of an asset over its useful life. Here are some of the most widely used methods:

A. Straight-Line Depreciation:
- Formula: Depreciation Expense =
                    (Cost of Asset-Residual Value) /Useful Life

- This method evenly spreads the cost of an asset over its useful life, resulting in a constant depreciation expense each year.

B. Declining Balance Depreciation:
- Formula: Depreciation Expense = Book Value at the Beginning of the Year × Depreciation Rate

- In this method, a fixed percentage is applied to the remaining book value of the asset each year, resulting in higher depreciation expenses in the earlier years.

C. Sum-of-Years-Digits (SYD) Depreciation:
- Formula: Depreciation Expense = (Remaining Useful Life\Sum of Years' Digits) × (Cost of Asset - Residual Value)

- SYD depreciation method accelerates depreciation by considering the sum of the digits representing the asset's useful life.

D. Units of Production Depreciation:
- Formula:
Depreciation Expense = (Number of Units Produced/Total Units Expected) × (Cost of Asset - Residual Value)

- This method ties depreciation to the actual usage or production output of the asset, making it suitable for assets where usage varies.

E. Double Declining Balance Depreciation:
- Formula: Depreciation Expense = 2 × (1/Useful Life) × (Cost of Asset - Accumulated Depreciation)

- Similar to declining balance, this method accelerates

depreciation, but it doubles the depreciation rate.

F. MACRS (Modified Accelerated Cost Recovery System):
- Commonly used for tax purposes, MACRS is a system defined by the Internal Revenue Service (IRS) with predefined depreciation schedules for different classes of assets.

G. Annuity Method:
- This method treats the depreciation as if it were an annuity, calculating the annual depreciation expense as a fixed percentage of the remaining book value.

Choosing the most appropriate method depends on factors such as the nature of the asset, its usage pattern, and the organization's financial objectives. Each method has its advantages and drawbacks, and businesses may use different methods for financial reporting, tax purposes, and management decisions.

This formula ensures a consistent distribution of depreciation expenses over the asset's useful life. It aligns accounting practices with economic realities and aids in accurate financial reporting.

14.3.2 Replacement Studies:
Replacement studies involve evaluating when and how assets should be replaced. The Total Cost of Ownership (TCO) formula is instrumental in this process:

$$TCO = \text{Purchase Cost} + \text{Operating Costs} + \text{Maintenance Costs} + \text{Disposal Costs}$$

Considering the TCO provides a comprehensive view of the costs associated with an asset throughout its lifecycle. A replacement decision is typically made when the TCO of a new asset is more favorable than the TCO of the existing one.

14.4 Conclusion
In conclusion, industrial costs are multifaceted and demand a holistic approach for effective management. Material cost control, labour cost control, and overhead cost control involve mathematical formulas that facilitate precise calculations and strategic decision-

making. Depreciation and replacement studies, driven by mathematical models, offer insights into the long-term planning required for maintaining a competitive edge. As businesses navigate the complexities of industrial costs, integrating technology, leveraging data analytics, and staying attuned to industry trends become imperative. The effective application of mathematical formulas in cost management is integral to achieving operational excellence, fostering innovation, and ensuring sustained growth in the dynamic business landscape.

Suggested Questions with Solutions

Material Cost Calculation

Question 1: ABC Manufacturing has an opening stock of raw materials valued at $50,000, makes purchases worth $120,000 during the year, and ends the year with a closing stock of $30,000. Calculate the material cost for the year.

Solution: Material Cost = Opening Stock + Purchases - Closing Stock
= $50,000 + $120,000 - $30,000 = $140,000

:

Labour Cost Control

Question 2: **XYZ Industries** pays its employees an average hourly wage of $20, and the total number of hours worked during the year is 50,000. Additionally, $5,000 is spent on overtime pay and $10,000 on employee benefits. Calculate the total labour cost.

Solution: Labour Cost = (Number of Hours Worked × Hourly Wage) + Overtime Pay + Benefits = (50,000 × $20) + $5,000 + $10,000 = $1,015,000

Overhead Cost Analysis

Question 3: The overhead costs for **LMN Corporation** are $80,000 for fixed overhead and $40,000 for variable overhead. Calculate the total overhead cost.

Solution: Total Overhead Cost = Fixed Overhead + Variable Overhead = $80,000 + $40,000 = $120,000

Straight-Line Depreciation

Question 4: A manufacturing company purchases a machine for $100,000 with a residual value of $10,000 and an estimated useful

life of 8 years. Calculate the annual straight-line depreciation.

Solution: Annual Depreciation = (Cost of Asset - Residual Value) \
Useful Life = ($100,000 - $10,000)\8 = $11,250 .

Replacement Studies

Question 5: A company is considering replacing an old machine. The existing machine has a purchase cost of $50,000, operating costs of $15,000 per year, maintenance costs of $5,000 per year, and disposal costs of $2,000. Calculate the Total Cost of Ownership (TCO) for the existing machine.

Solution: TCO = Purchase Cost + Operating Costs + Maintenance Costs + Disposal Costs = $50,000 + $15,000 + $5,000 + $2,000 = $72,000.

C Code for Depreciation and Replacement Analysis

```c
#include <stdio.h>

// Function to calculate straight-line depreciation
float calculateDepreciation(float cost, float residualValue, int usefulLife) {
    return (cost - residualValue) / usefulLife;
}

// Function to perform replacement analysis
void performReplacementAnalysis() {
    float existingTCO, newTCO;

    printf("Enter the Total Cost of Ownership (TCO) for the existing asset: $");
    scanf("%f", &existingTCO);

    printf("Enter the Total Cost of Ownership (TCO) for the potential replacement asset: $");
    scanf("%f", &newTCO);

    if (existingTCO < newTCO) {
        printf("Replacing the existing asset with the new asset is cost-effective.\n");
    } else if (existingTCO > newTCO) {
        printf("Keeping the existing asset is cost-effective.\n");
    } else {
        printf("The costs of the existing and potential replacement assets are equal.\n");
    }
}

int main() {
    float cost, residualValue, depreciation;
    int usefulLife, choice;

    // Input for asset details
    printf("Enter the cost of the asset: $");
```

```c
    scanf("%f", &cost);

    printf("Enter the residual value of the asset: $");
    scanf("%f", &residualValue);

    printf("Enter the useful life of the asset (in years): ");
    scanf("%d", &usefulLife);

    // Calculate and display straight-line depreciation
    depreciation = calculateDepreciation(cost, residualValue, usefulLife);
    printf("Straight-Line Depreciation: $%.2f per year\n", depreciation);

    // Ask the user if they want to perform replacement analysis
    printf("Do you want to perform a replacement analysis? (1 for Yes, 0 for No): ");
    scanf("%d", &choice);

    if (choice == 1) {
        performReplacementAnalysis();
    }

    return 0;
}
```

JAVA Code for Depreciation and Replacement Analysis

```java
import java.util.Scanner;

public class DepreciationAndReplacement {

    // Method to calculate straight-line depreciation
    private static float calculateDepreciation(float cost, float residualValue, int usefulLife) {
        return (cost - residualValue) / usefulLife;
    }

    // Method to perform replacement analysis
    private static void performReplacementAnalysis() {
        Scanner scanner = new Scanner(System.in);

        System.out.print("Enter the Total Cost of Ownership (TCO) for the existing asset: $");
        float existingTCO = scanner.nextFloat();

        System.out.print("Enter the Total Cost of Ownership (TCO) for the potential replacement asset: $");
        float newTCO = scanner.nextFloat();

        if (existingTCO < newTCO) {
            System.out.println("Replacing the existing asset with the new asset is cost-effective.");
        } else if (existingTCO > newTCO) {
            System.out.println("Keeping the existing asset is cost-effective.");
        } else {
            System.out.println("The costs of the existing and potential replacement assets are equal.");
        }
    }

    public static void main(String[] args) {
        Scanner scanner = new Scanner(System.in);
```

```
// Input for asset details
System.out.print("Enter the cost of the asset: $");
float cost = scanner.nextFloat();

System.out.print("Enter the residual value of the asset: $");
float residualValue = scanner.nextFloat();

System.out.print("Enter the useful life of the asset (in years):
");
int usefulLife = scanner.nextInt();

// Calculate and display straight-line depreciation
float depreciation = calculateDepreciation(cost, residualValue,
usefulLife);
System.out.printf("Straight-Line  Depreciation:  $%.2f  per
year\n", depreciation);

// Ask the user if they want to perform replacement analysis
System.out.print("Do you want to perform a replacement
analysis? (1 for Yes, 0 for No): ");
int choice = scanner.nextInt();

if (choice == 1) {
    performReplacementAnalysis();
}
}
}
```

PYTHON Code for Depreciation and Replacement Analysis

```python
def calculate_depreciation(cost, residual_value, useful_life):
    return (cost - residual_value) / useful_life

def perform_replacement_analysis():
    existing_tco = float(input("Enter the Total Cost of Ownership
(TCO) for the existing asset: $"))
    new_tco = float(input("Enter the Total Cost of Ownership
(TCO) for the potential replacement asset: $"))

    if existing_tco < new_tco:
        print("Replacing the existing asset with the new asset is cost-
effective.")
    elif existing_tco > new_tco:
        print("Keeping the existing asset is cost-effective.")
    else:
        print("The costs of the existing and potential replacement
assets are equal.")

def main():
    # Input for asset details
    cost = float(input("Enter the cost of the asset: $"))
    residual_value = float(input("Enter the residual value of the
asset: $"))
    useful_life = int(input("Enter the useful life of the asset (in
years): "))

    # Calculate and display straight-line depreciation
    depreciation = calculate_depreciation(cost, residual_value,
useful_life)
    print(f"Straight-Line Depreciation: ${depreciation:.2f} per
year")

    # Ask the user if they want to perform replacement analysis
    choice = int(input("Do you want to perform a replacement
analysis? (1 for Yes, 0 for No): "))
```

```python
    if choice == 1:
        perform_replacement_analysis()

if __name__ == "__main__":
    main()
```

15 FINANCIAL CONTROL: UNLOCKING OPERATIONAL EFFICIENCY AND STRATEGIC DECISION-MAKING

$\mathcal{F}$inancial control plays a pivotal role in industrial management, providing a structured approach to monitor, analyze, and optimize financial resources within an organization. This comprehensive exploration delves into key aspects of financial control, including ratio analysis, budgetary control, value analysis, and project evaluation methods such as payback period, discounted cash flow (DCF), and internal rate of return (IRR).

15.1 Ratio Analysis: A Window into Financial Health

Ratio analysis serves as a powerful tool in financial management, offering a comprehensive insight into the financial health and performance of an organization. These financial metrics act as a window, providing a clear view of various aspects of a company's operations. Liquidity ratios, such as the current ratio, reveal the company's ability to meet short-term obligations, offering a snapshot of its financial stability. Profitability ratios, such as gross and net profit margins, illuminate the company's ability to generate profits relative to its revenue. Efficiency ratios, exemplified by the asset turnover ratio, gauge how effectively the company utilizes its assets to generate sales. Solvency ratios, including the debt-to-equity ratio, shed light on the company's long-term financial viability. By scrutinizing these ratios, financial analysts and managers can identify trends, strengths, and weaknesses, enabling informed decision-making, strategic planning, and proactive measures to optimize financial performance and stability. Ratio analysis, therefore, stands as a fundamental tool in the financial toolkit, providing a nuanced and holistic understanding of an organization's financial landscape.

15.1.1 Liquidity Ratios
Liquidity ratios assess a company's ability to meet short-term obligations. The current ratio, calculated as current assets divided by current liabilities, offers insights into the firm's short-term solvency. A ratio above 1 indicates the company can cover its short-term liabilities.

$$\textbf{Current Ratio = Current Assets/Current Liabilities}$$

15.1.2 Profitability Ratios
Profitability ratios evaluate the company's ability to generate profits. The gross profit margin (gross profit divided by revenue) measures the efficiency of production, while the net profit margin (net profit divided by revenue) reflects overall profitability.

$$\textbf{Gross Profit Margin = Gross Profit/Revenue}$$

$$\textbf{Net Profit Margin = Net Profit/Revenue}$$

15.1.3 Efficiency Ratios
Efficiency ratios gauge how well a company utilizes its assets. The asset turnover ratio (revenue divided by average total assets) measures the efficiency of asset utilization in generating sales.

$$\textbf{Asset Turnover Ratio = Revenue/Average Total Assets}$$

15.1.4 Solvency Ratios
Solvency ratios assess a company's long-term financial viability. The

debt-to-equity ratio (total debt divided by shareholders' equity) indicates the proportion of financing provided by debt relative to equity.

Debt-to-Equity Ratio = Total Debt/Shareholders' Equity

Ratio analysis enables industrial managers to pinpoint areas of financial strength or weakness, guiding decisions related to capital structure, operational efficiency, and overall financial health.

15.2 Budgetary Control: Steering Financial Performance

Budgetary control functions as the rudder steering an organization's financial course, offering a systematic approach to planning, monitoring, and managing financial resources. This strategic tool plays a pivotal role in aligning financial performance with organizational goals. By setting comprehensive budgets encompassing various facets like sales, production, operating expenses, and capital expenditures, budgetary control establishes benchmarks for performance evaluation. The continuous monitoring of actual performance against these predetermined benchmarks through variance analysis enables financial managers to identify areas of strength or concern promptly. The process fosters cost discipline, efficient resource allocation, and the achievement of strategic objectives. Additionally, the flexibility and adaptability inherent in budgetary control ensure its relevance in dynamically changing business environments. The coordinated effort of creating, reviewing, and adjusting budgets not only facilitates efficient financial management but also enhances communication and coordination among different departments, creating a harmonious synergy toward achieving the organization's financial objectives. Ultimately, budgetary control emerges as a dynamic mechanism, guiding organizations in navigating their financial journey and steering toward sustained success.

15.2.1 Setting Budgets

Setting budgets is a fundamental aspect of financial management, serving as the foundation for strategic planning and control within organizations. This process involves meticulously outlining financial plans that delineate expected income, expenditures, and resource allocations over a specific period. Budgets are not merely numerical frameworks; they are reflections of an organization's priorities, goals, and operational strategies. Setting budgets requires a thorough understanding of past performance, market trends, and internal capabilities. It involves collaborative efforts across various departments to align financial objectives with overall organizational goals. Effective budget setting considers both short-term needs and long-term strategic visions. By establishing clear and realistic budgets, organizations provide a roadmap for financial decision-making, resource allocation, and performance evaluation. This proactive approach enables financial managers to anticipate challenges, identify opportunities, and maintain a disciplined control over financial resources, contributing to the overall success and sustainability of the organization.

15.2.2 Monitoring and Comparing Performance

Monitoring and comparing performance against established benchmarks is a critical phase in the budgetary control process, essential for maintaining financial health and achieving organizational goals. Once budgets are set, continuous scrutiny of actual performance becomes imperative. This ongoing monitoring allows financial managers to assess how well the organization is adhering to its financial plans and promptly identify any deviations or variances. The comparison of actual outcomes with budgeted figures serves as a diagnostic tool, highlighting areas of success and those requiring attention or corrective measures. This analytical approach not only provides insights into the effectiveness of resource allocation and expenditure but also enables proactive decision-making. Whether through regular financial reports or real-time analytics, monitoring and comparing performance create a feedback loop that fosters accountability, transparency, and a culture of continuous improvement within an organization. As a result, financial managers can make informed adjustments, align resources more effectively, and steer the organization toward optimal financial performance and strategic success.

15.2.3 Flexibility and Adaptability

Flexibility and adaptability are essential characteristics for any organization navigating the dynamic landscape of business and finance. In the realm of financial management, these qualities become particularly crucial in the context of budgetary control. A flexible approach implies the ability to adjust financial plans and strategies in response to changing circumstances, unforeseen challenges, or evolving market conditions. It acknowledges that the initial budget may need modification to stay relevant and effective. Adaptability, on the other hand, involves not only embracing change but also proactively seeking opportunities for improvement. A financial management system that is both flexible and adaptable ensures that budgets remain realistic and aligned with the organization's current needs and long-term objectives. This approach empowers financial managers to respond swiftly to emerging trends, technological advancements, or shifts in consumer behavior, thus enhancing the overall resilience and competitiveness of the organization in the ever-changing business environment.

Budgetary control not only fosters cost discipline but also enhances communication and coordination among different departments, aligning the entire organization towards common financial objectives.

15.3 Value Analysis: Enhancing Efficiency and Effectiveness

Value analysis stands as a strategic methodology aimed at enhancing the efficiency and effectiveness of organizational processes and products. Rooted in the principle of optimizing value while minimizing costs, value analysis delves into every aspect of a product or process to identify opportunities for improvement. This approach involves cross-functional collaboration, bringing together diverse perspectives from engineering, production, marketing, and finance to scrutinize each element of a product or service. By assessing the cost-value relationship, value analysis aims to eliminate non-value-added components, streamline operations, and enhance overall functionality. Furthermore, life cycle costing, a component of value analysis, considers the entire lifespan of a product, including production, operation, maintenance, and disposal costs, providing a holistic perspective. The value analysis process, therefore, not only promotes cost efficiency but also stimulates innovation, ensuring that resources are allocated to aspects that contribute the most significant value to the end-user. In essence, value analysis is a cornerstone of efficiency and effectiveness, aligning with industrial control principles and driving continuous improvement within an organization.

15.3.1 Cost-Value Relationship

The cost-value relationship is a fundamental concept in financial management and strategic decision-making. This relationship involves a careful examination of the costs associated with a product, service, or process relative to the perceived value it delivers to customers or end-users. The aim is to optimize this relationship by identifying and enhancing elements that contribute to value while minimizing unnecessary costs. Through methods like value analysis, organizations analyze each component of a product or process to discern where improvements can be made to increase overall value without inflating costs. This approach ensures that resources are allocated efficiently and that the end result aligns with customer expectations. The cost-value relationship is integral to maintaining a competitive edge, as it enables organizations to deliver high-quality products or services at optimal costs, fostering customer satisfaction and long-term success.

15.3.2 Cross-Functional Collaboration

Cross-functional collaboration is a strategic approach that involves breaking down traditional silos within an organization and fostering teamwork among individuals from different departments or disciplines. This collaborative framework encourages the sharing of diverse perspectives, knowledge, and skills to achieve common goals. In the context of industrial and financial management, cross-functional collaboration is crucial for addressing complex challenges that require a multidimensional approach. Teams comprising members from various functions, such as engineering, production, marketing, and finance, work collectively to analyze problems, identify opportunities, and implement solutions. This collaborative synergy not only enhances communication and understanding across departments but also promotes innovation and efficiency. In financial management, cross-functional collaboration is particularly valuable during processes such as budgeting, value analysis, and strategic planning, where insights from multiple disciplines contribute to more robust and well-rounded decision-making. Ultimately, cross-functional collaboration is a catalyst for organizational success, fostering a culture of cooperation, adaptability, and collective achievement.

15.3.3 Life Cycle Costing

Life Cycle Costing (LCC) is a comprehensive and forward-thinking financial approach that considers the entire lifespan of a product, service, or project. It goes beyond traditional cost accounting methods by incorporating not only the initial acquisition or production costs but also the costs associated with its operation, maintenance, and eventual disposal. In essence, LCC provides a holistic perspective on the financial implications of an asset over its entire life. This approach enables organizations to make more informed decisions by factoring in long-term considerations and potential future expenses. For instance, in industrial settings, Life Cycle Costing can guide decisions regarding equipment purchases, taking into account not just the upfront costs but also the expected maintenance expenses and energy consumption over its operational life. Embracing Life Cycle Costing enhances strategic planning, as it encourages a more sustainable and economically sound use of resources, aligning financial decisions with the organization's long-term goals and environmental considerations.

Value analysis aligns with the principles of industrial control by optimizing resources, enhancing efficiency, and maintaining a focus on delivering maximum value to stakeholders.

15.4 Project Evaluation: Navigating Investment Decisions

Project evaluation is a critical process in financial management that serves as a compass for navigating investment decisions within an organization. It involves a systematic assessment of potential projects to determine their viability, profitability, and alignment with strategic goals. Three key methodologies used in project evaluation include the Payback Period, Discounted Cash Flow (DCF), and Internal Rate of Return (IRR). The Payback Period calculates the time required for an investment to recover its initial cost, offering a quick assessment of liquidity. Discounted Cash Flow adjusts future cash inflows to their present value, considering the time value of money, and Net Present Value (NPV) is a crucial metric derived from this method. Meanwhile, Internal Rate of Return represents the discount rate that makes the NPV of a project equal to zero, providing insights into the project's rate of return. These evaluation techniques empower decision-makers to make informed choices, ensuring that investment decisions align with the organization's financial capabilities, risk tolerance, and strategic objectives. Project evaluation, therefore, is a navigational tool guiding organizations through the complex landscape of investment decisions, helping them capitalize on opportunities and mitigate risks. This section explores three key evaluation techniques: payback period, discounted cash flow (DCF), and internal rate of return (IRR).

15.4.1 Payback Period

The payback period represents the time required for an investment to recoup its initial cost. This method is straightforward but has limitations, as it doesn't consider the time value of money.

Payback Period = Initial Investment/Annual Cash Inflow

15.4.2 Discounted Cash Flow (DCF)

DCF adjusts future cash flows to their present value, considering the time value of money. Net present value (NPV) is a common DCF metric, representing the difference between the present value of cash inflows and outflows.

$$NPV = \sum_{t=0}^{n} \frac{Cash - flow_t}{(1 + r)^t} - \text{Initial Investment}$$

Where:
- Cash-flow is the cash flow in year t.
- r is the discount rate.
- n is the project's useful life.

15.4.3 Internal Rate of Return (IRR)

IRR is the discount rate that makes the net present value of a project equal to zero. It represents the project's rate of return, indicating its attractiveness.

IRR : NPV = 0

Project evaluation methods guide industrial managers in selecting projects with the most favorable financial outcomes, aligning investments with organizational goals and financial capabilities.

15.5 Conclusion: Integrated Financial Control for Industrial Excellence

In conclusion, financial control, comprising ratio analysis, budgetary control, value analysis, and project evaluation, serves as the backbone of industrial management. Ratio analysis unveils the financial health of an organization, providing insights into liquidity, profitability, efficiency, and solvency. Budgetary control ensures disciplined financial management, aligning organizational efforts with predefined financial objectives.

Value analysis promotes efficiency and effectiveness by optimizing costs and enhancing value delivery. Project evaluation methods, such as payback period, discounted cash flow, and internal rate of return, empower decision-makers to make informed choices regarding investments, ensuring alignment with strategic goals.

Industrial managers must adopt an integrated approach to financial control, recognizing its multifaceted role in steering the organization towards operational excellence, strategic success, and sustained financial health. By embracing these financial control mechanisms, industrial enterprises can navigate complex financial landscapes, mitigate risks, and capitalize on opportunities, thereby fostering a robust and adaptive financial framework for long-term prosperity.

Suggested Questions with Solutions

Ratio Analysis

Question 1: ABC Company has a current ratio of 2.5 and a quick ratio of 1.8. If the current liabilities are $80,000, calculate the current assets and the quick assets.

Solution:
Current Assets = Current Ratio × Current Liabilities
Quick Assets = Quick Ratio × Current Liabilities

Substitute the given values:
Current Assets = 2.5 × $80,000 = $200,000
Quick Assets = 1.8 × $80,000 = $144,000

Budgetary Control

Question 2: XYZ Corporation sets a budget of $500,000 for production expenses. If actual expenses amount to $480,000, calculate the variance and comment on the company's performance.

Solution:
Variance = Budgeted Expenses-Actual Expenses
= $500,000 - $480,000 = $20,000

Comment: The company has a favorable variance of $20,000, indicating that actual production expenses were lower than budgeted, reflecting cost efficiency.

Value Analysis

Question 3: An organization adopts value analysis to assess a manufacturing process costing $3,50,000. If the analysis identifies cost-saving opportunities worth $50,000 without compromising quality, calculate the new total cost and comment

on the impact.

Solution:
New Total Cost = Old Total Cost - Cost Savings
=$3,50,000- $50,000 = $3,00,000

Comment: The organization's total cost decreases by $50,000, showcasing the positive impact of value analysis on cost optimization without compromising quality.

Payback Period

Question 4: Company P invests $200,000 in a new project, and it generates an annual cash inflow of $60,000. Calculate the payback period for the project.

Solution:
Payback Period = Initial Investment\Annual Cash Inflow
= $200,000\$60,000 =3.33 years

Internal Rate of Return (IRR)

Question 5: An investment project has an initial cost of $150,000 and generates annual cash flows of $30,000 for five years. Calculate the IRR for the project.

Solution:

IRR : NPV = 0

Use the discounted cash flow formula:

$$NPV = \sum_{t=0}^{n} \frac{Cash - flow_t}{(1+r)^t} - \text{Initial Investment}$$

Solve for IRR:
IRR $\approx 0.12 = 12\%$

C Code for Calculating Payback Period and IRR

```c
#include <stdio.h>
#include <math.h>

// Function to calculate payback period
float calculatePaybackPeriod(float initialInvestment, float annualCashInflow) {
    return initialInvestment / annualCashInflow;
}

// Function to calculate internal rate of return (IRR) using the Newton-Raphson method
float calculateIRR(float initialInvestment, float annualCashInflows[], int projectLife) {
    float irr = 0.1; // Initial guess for IRR
    float epsilon = 0.001; // Tolerance for convergence
    float npv;

    do {
        npv = -initialInvestment; // Initial NPV at the guessed IRR

        for (int t = 0; t < projectLife; ++t) {
            npv += annualCashInflows[t] / pow(1 + irr, t + 1);
        }

        // Derivative of NPV with respect to IRR
        float npvDerivative = 0.0;

        for (int t = 0; t < projectLife; ++t) {
            npvDerivative += -t * annualCashInflows[t] / pow(1 + irr, t + 2);
        }

        // Update IRR using the Newton-Raphson formula
        irr -= npv / npvDerivative;

    } while (fabs(npv) > epsilon);
```

```
        return irr;
    }

    int main() {
        float initialInvestment = 150000.0;
        float annualCashInflows[] = {30000.0, 30000.0, 30000.0,
30000.0, 30000.0}; // Example annual cash inflows
        int     projectLife    =     sizeof(annualCashInflows)    /
sizeof(annualCashInflows[0]);

        // Calculate and display payback period
        float paybackPeriod = calculatePaybackPeriod(initialInvestment,
annualCashInflows[0]);
        printf("Payback Period: %.2f years\n", paybackPeriod);

        // Calculate and display internal rate of return (IRR)
        float irr = calculateIRR(initialInvestment, annualCashInflows,
projectLife);
        printf("Internal Rate of Return (IRR): %.2f%%\n", irr * 100);

        return 0;
    }
```

JAVA Code for Calculating Payback Period and IRR

```java
import java.util.Arrays;

public class InvestmentAnalysis {

    // Function to calculate payback period
    private static double calculatePaybackPeriod(double
initialInvestment, double annualCashInflow) {
        return initialInvestment / annualCashInflow;
    }

    // Function to calculate internal rate of return (IRR)
    private static double calculateIRR(double initialInvestment,
double[] annualCashInflows) {
        double irr = 0.1; // Initial guess for IRR
        double epsilon = 0.001; // Tolerance for convergence
        double npv;

        do {
            npv = -initialInvestment;  // Initial NPV at the guessed
IRR

            for (int t = 0; t < annualCashInflows.length; ++t) {
                npv += annualCashInflows[t] / Math.pow(1 + irr, t +
1);
            }

            // Derivative of NPV with respect to IRR
            double npvDerivative = 0.0;

            for (int t = 0; t < annualCashInflows.length; ++t) {
                npvDerivative += -t * annualCashInflows[t] /
Math.pow(1 + irr, t + 2);
            }

            // Update IRR using the Newton-Raphson formula
            irr -= npv / npvDerivative;
```

```
        } while (Math.abs(npv) > epsilon);

        return irr;
    }

    public static void main(String[] args) {
        double initialInvestment = 150000.0;
        double[] annualCashInflows = {30000.0, 30000.0, 30000.0,
30000.0, 30000.0}; // Example annual cash inflows

        // Calculate and display payback period
        double                    paybackPeriod              =
calculatePaybackPeriod(initialInvestment, annualCashInflows[0]);
        System.out.printf("Payback    Period:    %.2f    years\n",
paybackPeriod);

        // Calculate and display internal rate of return (IRR)
        double    irr    =    calculateIRR(initialInvestment,
annualCashInflows);
        System.out.printf("Internal    Rate    of    Return    (IRR):
%.2f%%\n", irr * 100);
    }
}
```

PYTHON Code for Calculating Payback Period and IRR

```python
def                      calculate_payback_period(initial_investment,
annual_cash_inflow):
    return initial_investment / annual_cash_inflow

def calculate_irr(initial_investment, annual_cash_inflows):
    irr = 0.1  # Initial guess for IRR
    epsilon = 0.001  # Tolerance for convergence
    npv = 0.0

    while True:
        npv = -initial_investment  # Initial NPV at the guessed IRR

        for t, cash_flow in enumerate(annual_cash_inflows):
            npv += cash_flow / (1 + irr) ** (t + 1)

        # Derivative of NPV with respect to IRR
        npv_derivative = 0.0

        for t, cash_flow in enumerate(annual_cash_inflows):
            npv_derivative += -t * cash_flow / (1 + irr) ** (t + 2)

        # Update IRR using the Newton-Raphson formula
        irr -= npv / npv_derivative

        if abs(npv) < epsilon:
            break

    return irr

def main():
    initial_investment = 150000.0
    annual_cash_inflows  =  [30000.0,  30000.0,  30000.0,  30000.0,
30000.0]  # Example annual cash inflows

    # Calculate and display payback period
    payback_period = calculate_payback_period(initial_investment,
annual_cash_inflows[0])
```

```python
    print(f"Payback Period: {payback_period:.2f} years")

    # Calculate and display internal rate of return (IRR)
    irr = calculate_irr(initial_investment, annual_cash_inflows)
    print(f"Internal Rate of Return (IRR): {irr * 100:.2f}%")

if __name__ == "__main__":
    main()
```

ABOUT THE AUTHOR

Dr. Prasun Bhattacharjee is a Ph.D. in Engineering (Awarded by the Department of Mechanical Engineering of the prestigious Jadavpur University of Kolkata, India). His numerous scientific contributions have been published in distinguished peer-reviewed journals. Prasun has also presented his research works at several international conferences held in the USA and European nations. He is currently a member of the Indian Institute of Welding and the Association for Information Systems. His research mainly focuses on employing artificial intelligence techniques to enhance the performance of wind power generation systems. Dr. Bhattacharjee earned the university medal of the Maulana Abul Kalam Azad University of Technology, West Bengal while studying for the Master of Technology degree in Industrial Engineering and Management. He has also worked for the distinguished TATA group as a Systems Engineer after passing out from the Kalyani Government Engineering College as a mechanical engineer. Dr. Bhattacharjee has traveled extensively to almost every corner of India and 25 foreign nations with his parents. He loves to share his travel experiences with other fellow nomads to help them witness the wonders of the world on their own. You can enjoy his exhilarating travel videos on his YouTube channel (https://www.youtube.com/c/prasunbhattacharjee1206) or visit the author on Twitter (@Prasun6official). Dr. Bhattacharjee is presently serving as a faculty in Mechanical Engineering.